1. Introduction

As related in the following statements from the General der Artillerie beim Chef General Stab des Heeres, the Jagdpanzer IV was actually an invention created by the artillery as a newly designed Sturmgeschütz. On 2 February 1944 General Lindemann (General der Artillerie) sent the following comments on the proposal by General Guderian (Generalinspekteurs der Panzertruppen) that the Sturmgeschütz n.A. be renamed as a 'Panzerjäger'.

The Sturmgeschütz n.A. is a technically advanced development of the old Sturmgeschütz. Based on Sturmartillerie combat experience, the requirements for the new design of the Sturmgeschütz were sent to the AHA/In4 and Waffenamt by the Sturmartillerie Lehr-Abteilung on 5 October 1942, which was months before the creation of the Gen.Insp.d.Pz.Tr.

Based on this technical development series and the fact that the Sturmgeschütz in knocking out over 13,000 enemy tanks is the most effective armoured vehicle with the infantry, we are against renaming Sturmartillerie's further development of the Sturmgeschütz.

Because Sturmgeschütz fire 25% of their ammunition at tanks and 75% at other types of targets, the designation 'Panzerjäger' relates to only part of the Sturmgeschütz' assigned tasks.

The designation 'Sturmgeschütz' is a well known concept for the infantry. Therefore the General der Infanterie is for retaining the Sturmgeschütz designation.

It is a sure supposition that with increased production the Sturmgeschütz n.A. will also be used by the Sturmartillerie. So that it doesn't appear to be suitable to name a weapon as a 'Panzerjäger' when it is used by both the Sturmartillerie and Panzerjäger units.

The actual point being argued went far beyond a name. Control of the issue and tactical employment of the Sturmgeschütz n.A. was at stake. The Gen.Insp.d.Pz.Tr., controlled issue of Panzerjäger but only part of the Sturmgeschütz. General Guderian wanted to issue this new weapon solely to Panzerjäger units. The reality was that the Panzerjäger units assigned to Panzer-Divisions only had previous experience in using lightly armoured self-propelled anti-tank guns, which were tactically employed quite differently than a fully armoured 'Sturmgeschütz' which could withstand hits from enemy tanks.

As it turned out, Guderian got his way. The Jagdpanzer IV along with the upgunned Panzer IV/70 (V) were only issued to inexperienced Panzerjäger units. They were never issued to Sturmartillerie units which had perfected counter strike tactics to achieve a very high kill ratio against enemy tanks. Allied tankers should be very grateful that these dangerous anti-tank weapons were given to novices instead of the proven experts at knocking out tanks.

Many days were spent in exactly measuring all of the external features on surviving Jagdpanzer IV, Panzer IV/70 (V), and Panzer IV/70 (A) in order to accurately create for the first time the as-built drawings featured in this Panzer Tracts. Those measured in detail included the:

Versuchs-Jagdpanzer IV (Fgst.Nr.V2) in Germany
Jagdpanzer IV (60mm) (Aufbau Nr.25271) in France
Jagdpanzer IV (80mm) (Aufbau Nr.25476) in Germany
Panzer IV/70 (V) (Fgst.Nr.320864) in USA
Panzer IV/70 (V) (Fgst.Nr.320996) in USA
Panzer IV/70 (V) (Fgst.Nr.329667) in Canada
Panzer IV/70 (A) (Fgst.Nr.120539) in France

Because details are hidden on consolidated drawings of the entire Panzer at 1:35 scale, drawings of the individual parts are printed at large 1:10 scale on separate pages. There is no loss in detail with reduction in scale, because the software now being used to create the printing plates draws the lines, circles, arcs, hexes, and ovals one at a time - just like we did in creating these accurate as-built drawings.

The most significant discovery was counter intuitive. The superstructure roof length is 1890mm +/- 2 on a Jagdpanzer IV (regardless of whether the frontal armour is 60mm or 80mm thick). With the installation of the longer 7.5cm Pak 42 L/70 to create the Panzer IV/70 (V), one would assume that the length of the superstructure would either remain the same or be lengthened. Actually, the length of superstructure roof on a Panzer IV/70 (V) was shortened by 30mm to 1860mm. While the superstructure roof on the Panzer IV/70 (A) is shorter still at 1847mm.

It was also remarkable to discover that there were over 10 changes between Panzer IV/70 (V) (Fgst.Nr.320864) completed in October 1944 and Panzer IV/70 (V) (Fgst.Nr.329667) completed in March 1945, including the front track guards, air vents on brake access hatches, bump stops for the machinegun armour cover, external travel lock, tabs for securing the tarp cover, curved armour guard for the periscopic gunsight, rain channels for the roof hatches, pegs to mount a rangefinder, battle sight for the commander, chains to raise the cooling air intake/outlet flaps, muffler, and centred tow coupling. It is astounding that during this late-war period of dire circumstances, engineers were allowed to introduce insignificant cosmetic changes which did not improve combat effectiveness but did interfere with mass production.

As is our high standard, Panzer Tracts are based solely on surviving specimens, wartime photographs, and the content of primary source documents written by those who participated in the design, production, and employment of the Panzers. Without injecting postwar opinions, conjecture, and speculation, we allow the real experts who designed, produced, and used the Jagdpanzer IV, Panzer IV/70 (V) and Panzer IV/70 (A) to have their say.

Left: The full scale wooden model for the superstructure of the kleine Panzerjäger der Firma Vomag was mounted on a normal Pz.Kpfw.IV Ausf.F chassis. (TTM)

Below: Two machinegun ports, with associated cast armour guards were added to the superstructure front before it was displayed to Hitler on 20 October 1943. (BSB)

2. Jagdpanzer IV

Panzerjäger IV (7.5cm Pak 39 L/48) (Sd.Kfz.162) Ausf.F. Fgst.Nr. Serie 320001 - 321000

2.1 Development

Due to the tremendous success of the Sturmgeschütz in knocking out enemy tanks and the desire to unify chassis types, the firm Vomag was given an order by the Waffenamt in September 1942 to proceed with the detailed design of a Sturmgeschütz mounting a 7.5cm gun on a modified Pz.Kpfw.IV chassis. A wooden model of the superstructure was completed and displayed for Hitler at a conference held 13 May 1943. This preliminary design was based on the normal Pz.Kpfw.IV chassis with the 7.5cm gun mounted in a low superstructure. Striving for a low silhouette this initial design had only an overall height of 1.7 metres. While a low silhouette is preferred, this height was considered to be at the limit of being too low for combat in rolling terrain. Additional excellent features were the well sloped superstructure frontal armour and the gun mantle specifically designed to prevent shot traps.

During the Summer of 1943 improvements to the design continued. The most important feature was redesigning the layout of the hull frontal armour. The basic Pz.Kpfw.IV chassis frontal armour had evolved from 15mm to 80mm mounted at an angle to the vertical of only 15°. While 80mm at this angle was good enough to prevent penetration at normal combat ranges by the 75mm gun on the Sherman and the 76.2mm gun on the Russian T34, it was not effective when engaged by the British 17-Pounder tank gun or the Russian 85mm tank gun.

As early as February 1943 modifications had been proposed by the Heeres Waffenamt to increase the frontal armour of the Pz.Kpfw.IV by increasing the angle to the vertical. These new designs were turned down each and every time by both the armour manufacturers and the assembly firms for the same basic reason that this extensive modification would have disrupted production when every Pz.Kpfw.IV was sorely needed at the front. But, with an entirely new design for the Jagdpanzer IV, the changes could be easily initiated prior to and without interrupting production. This new front hull design had an upper plate of 60mm at 45° and a lower plate of 50mm at 55° which gave the equivalent protection of rolled armour plate at 0° of 110mm and 123mm respectively.

A trial leichter Panzerjäger mit 7.5cm L/48 auf Fahrgestell IV (Eisen) constructed out of soft steel was demonstrated at Hitler's conference on 20 October 1943. Two machinegun ports, mounted on the right and left of the main gun mount in the superstructure front, had been added. The Versuchs-Jagdpanzer IV had rounded corners on the superstructure which were deleted from series production vehicles since bending the rolled plate would have added unnecessary time and labour for the armour manufacturer without adding significantly to the armour protection. Circular plugged ports (MP Stopfen) in the superstructure sides that were present on the trial hull were also dropped from production vehicles since a close-defence weapon, the Nahverteidigungswaffe, was mounted on the superstructure roof of production vehicles.

As with all armoured vehicles developed and produced by Germany in World War II, the Jagdpanzer IV was known by a series of unofficial descriptive names as follows:

- **kleine Panzerjäger der Firma Vomag**
  May43
- **Panzerjäger auf Fahrgestell Panzer IV**
  Aug43
- **leichter Panzerjäger mit 7.5cm L/48 auf Fgst.IV**
  Sep43
- **Stu.Gesch.n.A. (le.Pz.Jg.) auf Pz.IV**
  Dec43 as well as the official titles:
- **leichter Panzerjäger auf Fgst.Pz.Kpf.Wg.IV mit 7.5cm Pak 39 L/48**
  Wa J Rü Oct43 to Dec43
- **le.Pz.Jg.IV instead of Sturmgeschütz neuer Art** Wa Prüf 28Jan44
- **Panzerjäger IV (7.5cm Pak 39 (L/48) (Sd.Kfz.162)**
  K.St.N.1149 1Feb44
- **Panzerjäger IV (Sd.Kfz.162)**
  Vomag Mar44
- **Sturmgeschütz neuer Art mit 7.5cm Pak 39 L/48 auf Fgst. Pz.Kpfw.IV**
  Wa J Rü Feb44 to Oct44
- **le.Pz.Jg.Vomag mit 7.5cm Kw.K. L/48 auf Fgst.Pz.IV als 'le. Panzerjäger IV' (ehem. Stu.Gesch.n.A.)**
  Gen.Insp.d.Pz.Tr. 8Aug44
- **Jagdpanzer IV Ausf.F**
  D 653/39 15Sep44
- **Jagdpanzer IV - Panzerjäger IV (m.7.5cm Pak 39 L/48) (Sd.Kfz.162)**
  Wa J Rü from Nov44

2.2 Characteristics

The basic design of the chassis, suspension and drive train for the Jagdpanzer IV was adopted from the Pz.Kpfw. IV Ausf.F. Components that remained basically unaltered were the Maybach HL 120 TRM engine, the Zahnradfabrik SSG76 six-speed synchromesh transmission, the steering and final drive assemblies, the motor compartment, rear deck, hull rear and all suspension components including the drive sprocket, idler, return rollers, roadwheels and track. The basic features that were altered were the frontal armour, including a shortened glacis plate set at a greater angle, the drive shaft and the fuel tanks. In the basic Pz.Kpfw.IV the two main fuel tanks were located under the turret floor with a third small fuel tank under the left rear deck for the auxiliary generator engine. But, in order to lower the vehicle profile, the fuel tank arrangement had to be changed. In the Jagdpanzer IV, two forward fuel tanks were mounted under the gun across the hull with a third fuel tank in the lower left engine compartment. To fill these tanks, two armoured ports were located in the left hull side. The last significant change to the hull was relocating the escape hatch. Originally located under the radio operator's position in the Pz.Kpfw.IV, this position was not readily accessible in the Jagdpanzer IV, therefore, it was relocated to a centre left position in the hull floor underneath the gunner's position in the Jagdpanzer IV.

While much of the hull was the same basic design as its parent, the superstructure of the Jagdpanzer IV was a completely new design. All plates making up the superstructure sides were well sloped with the front 60mm plate at 50°, the 40mm side plates at 30° and the 30mm rear plate at 33°. The gun mount was bolted to the superstructure front plate about 20cm to the right of the centre line because of the gunsight. This gun mount for the 7.5cm Pak 39 L/48 consisted of a large stationary casting with an inner ball shaped shield and an outer Topfblende (cast gun mantlet) which provided equivalent or greater protection than the sloped frontal armour. Additional fire power was provided by a machinegun that could be mounted in one of two ports in the superstructure front, a Rundumfeuer machinegun mounted on the superstructure roof that could be rotated 360° and fired remotely from inside the vehicle, and a Nahverteidigungswaffe that could be rotated 360°. This Nahverteidigungswaffe (close defence weapon) could fire either smoke or high explosive projectiles in order to screen the vehicle or discourage close in attack by infantry tank hunter teams. Seventy-nine rounds of ammunition for the main gun were carried along with 1200 rounds for the machinegun, 384 rounds for the machine pistols and 16 rounds for the Nahverteidigungswaffe. Normally half of the ammunition for the main gun was high explosive and half was armour-piercing. But, the type of ammunition carried often varied depending on the action expected and the ammunition available.

A crew of four manned the Jagdpanzer IV. The driver seated in the left front did not have a separate hatch. His vision for driving was provided by two slits with glass episcopes in the superstructure front plate. The gunner seated directly behind the driver had a periscopic dial sight Sfl.ZF1 that extended through an arc shaped opening in the superstructure roof. The commander located behind the gunner had a hatch with two covers. The forward hatch could be opened to raise the Sf.14Z scissors periscope for observation while the rear hatch cover remained closed. The commander would also use a swivelling periscope mounted in the larger rear hatch or a stationary periscope facing nine o'clock that was mounted to his left in the superstructure roof. The loader who also served as the radio operator was the only crew member on the right side of the vehicle. He had a sighting aperture for the machine gun port in the superstructure front and a stationary periscope facing three o'clock in the superstructure roof. A circular hatch was mounted in the superstructure roof above his position

2.3 Production

The sole armour supplier for the Pz.Jg.IV was Witkowitz (code 'bzs') which reported completion of 46 Wanne (armour hulls) and 25 Aufbau (superstructures) by the end of 1943. A total of 405 Wanne (104 in Jan, 75 in Feb, 62 in Mar, 66 in Apr, 98 in May) and 403 Aufbau (90 in Jan, 49 in Feb, 62 in Mar, 89 in April, 113 in May) were completed by the end of May 1944 with production scheduled to increase from 120 in June to 180 per month by November 1944. Witkowitz stamped their production code 'bzs' and a five digit serial number beginning with 25001 into their completed armour components.

As planned in July 1943, the first 10 le.Pz.Jg. were to be completed by Vomag (code 'ajk') in September 1943 with production gradually increasing by 10 per month to 20 in October, 30 in November, and 40 in December.

However, problems had delayed production as reported in a meeting with HDL Saur on 9 January 1944:

A total of 10 le.Pz.Jg. were completed by Vomag in December 1943. Vomag production was uncertain because the gun mount was not in order and there were problems with the quality of the steel castings from Witkowitz. Vomag's director stated that output would be 30 le.Pz.Jg. by the end of January, including the 10 completed in December.

As reported by Wa J Rü, the first 30 Jagdpanzer IV were accepted by the Waffenamt in January 1944. At this time Vomag was still producing Pz.Kpfw.IV which it continued through May 1944. The production of the Pz.Kpfw.IV was phased out as the production of the Jagdpanzer IV increased. As shown on the accompanying table of production and issue, production steadily increased to over 100 in April and reached a peak of 125 in July 1944. With the introduction of the longer 7.5cm Pak L/70 in August, the Jagdpanzer IV with its shorter 7.5cm Pak39 L/48 was phased out.

Having missed the production goal of 140, in May 1944 Wa J Rü reported that in addition to the 90 accepted by the inspectors, another 26 Fahrgestell had been completed. Again having missed the goal of 130 in July 1944, Wa J Rü reported that in additional to the 125 accepted by inspectors another 15 had been completed but still had not been accepted. The correct total completed and accepted by inspectors of 90 in May and 125 in July is confirmed by reports independently created by the Hauptauschuss Panzerwagen.

Vomag was hit by a heavy bombing raid that severely disrupting production in September 1944. The initial report that Vomag had completed 30 le.Pz.Jg. 7.5cm L/48 in September 1944 was corrected to 19 in a Hauptausschuss Panzerwagen report dated 7 November 1944. A further 46 Jagdpanzer IV were completed in October and the last two were in November 1944 for a total of 750 Jagdpanzer IV. This total of 750 is confirmed by totalling the numbers in the new production released column on the monthly Wa J Rü reports which add up to 751.

2.3.1 Production Figures

Month	Planned	Accepted
Jan44	50	30
Feb44	60	45
Mar44	90	75
Apr44	120	106
May44	140	90
Jun44	120	120
Jul44	130	125
Aug44	80	92
Sep44	60	19
Oct44	54	46
Nov44	0	2
Total	904	750

2.4 Modifications Introduced During Production

As with all German armoured vehicles that remained in production for an extended period, there was no such item as a 'standard' Jagdpanzer IV. Modifications were continuously introduced to improve the characteristics and simplify production. Some of the changes were forced due to materials or parts shortages. During the production run of the Jagdpanzer IV, the following significant changes were made:

2.4.1 January 1944

All Jagdpanzer IV had a circular hole cut in the roof for the Nahverteidigungswaffe. Due to a production shortage, this weapon was frequently unavailable for installation. To cover the hole most Jagdpanzer IV had a circular armoured cover held in place by four bolts.

2.4.2 February 1944

In order to lighten the front, the spare track originally mounted across the front upper hull was moved to the upper hull rear. The two spare roadwheels that had been mounted on the upper hull rear were moved to mounts on the left engine hatch on the rear deck.

2.4.3 March 1944

The left machinegun port was deleted since it could not be effectively utilised. Since this port was cut by the armour manufacturer, the assembly firm welded a 60mm thick circular bevelled cap over the opening until superstructures were delivered without the port cut into the armour plate. A Rundumfeuer machinegun was mounted on the superstructure roof in front of the loader's hatch on a trial basis on several Jagdpanzer IV produced in March and April 1944. As recorded at a demonstration on 8 April 1944, there was a problem with sufficient space to mount the Rundumfeuer M.G.

2.4.4 April 1944

The lower corners were cut from the inner gun mantle base plate on the superstructure in order to further reduce the weight of this front heavy vehicle. As recorded at a demonstration on 8 April 1944, the 7.5cm Pak 39 no longer had a muzzle brake. The muzzle brake was not needed when continuously firing up to 50 rounds. An improved Rohrbremse (recoil cylinder) was to be installed in future production. The troops removed the muzzle brakes from most of the vehicles that had already been issued.

As announced in the H.T.V.Bl. (army technical orders pamphlet) by In 6 on 29 August 1944:

The 7.5cm Pak 39 (L/48) was issued to the troops in two models. The alte Ausführung (old model) had a 5 litre Rohrbremse (recoil cylinder) with a 52mm diameter rod and a hydraulic safety switch. The Mundungsbremse (muzzle brake) for this old model only had the purpose of preventing the recoil cylinder from overheating when the gun was continuously fired. The muzzle brake was dropped because the old model recoil cylinder could take the recoil forces and stresses from continuous firing. The neue Ausführung had a 6 litre Rohrbremse with a 60mm diameter piston rod and without a hydraulic safety switch. In both models, the air pressure in the Luftvorholer (recuperator) was to be 45 +/- 3 kg/cm^2.

2.4.5 May 1944

Starting with Fahrgestell Nr.320301, the armour thickness of the superstructure and upper hull front plates was increased to 80mm (confirmed by examination of Fgst. Nr.320314). The diameter of the conical cover over the superstructure machinegun port was increased and a notch was cut out of the right side of the inner gun mount base plate to allow space for the larger conical armour cover to be pivoted.

2.4.6 June 1944

In order to simplify construction, the sides of the armoured box over the radiator filler caps on the rear deck were cut square instead of at the previous sloped angle.

2.4.7 September 1944

Due to rumours that the anti-magnetic coating Zimmerit caught fire due to hits from anti-tank rounds even when they did not penetrate, Zimmerit was no longer applied to Jagdpanzer IV at the factory. After complaints from the troops of rain leaking into the vehicle, small brackets were welded to the superstructure front and sides near the top and at the rear of the superstructure roof to anchor straps for a Regenplane (rain tarpaulin) to cover the entire superstructure roof.

2.4.8 October 1944

In order to save on the scarce roller bearings and to reduce manufacturing time, the number of return rollers on each side was reduced from four to three. In addition to these modifications, command versions were produced for the Jagdpanzer IV issued to the Abteilung headquarters and the Jagdpanzer IV issued to each company commander. These command versions are externally identifiable by a second antenna mount at the left front corner of the rear deck. This second antenna was for the longer range FuG8 radio set in command vehicles which was carried along with the normal FuG5 radio set A fifth crew member was added as a radio operator in this command version.

2.5 Combat Service

Starting in March 1944, Jagdpanzer IV were issued to the Panzer-Jäger-Abteilungen of Panzer Divisions and Panzer Grenadier Divisions. Each Panzer Jäger Kompanie were issued either 10 or 14 Jagdpanzer IV in accordance with the organization outlined in K.St.N. 1149. In most cases the Panzer Jäger Abteilung in the Panzer Divisions had two companies each with 10 Jagdpanzer IV and one Jagdpanzer IV for the Abteilung commander. The Panzer Jäger Abteilung in the Panzer-Grenadier Divisions had two companies each with 14 Jagdpanzer IV along with 3 Jagdpanzer IV in the Abteilung headquarters section.

There were a few exceptions to this general rule. Panzer Jäger Lehr Abteilung 130 in the Panzer-Lehr-Division was the first unit to receive Jagdpanzer IV. The original plan was to outfit only one company with 14 Jagdpanzer IV and another company with 14 Jagdtiger. But, due to production delays, the Jagdtiger were never issued to this unit, resulting in the Pz.Jg.Lehr Abt.130 reorganising so that there were 9 Jagdpanzer IV with each of the three companies and 4 Jagdpanzer IV with the Abteilung headquarters. A second exception was the Fallschirm-Panzer-Division 'Hermann Göring' which employed the Jagdpanzer IV issued in April 1944 in the III.Abteilung of their Panzer Regiment. The second time Jagdpanzer IV were issued to this division, 10 were given to each of the three companies and one given to the commander of the Panzer-Jäger-Abteilung 'Hermann Göring'. The balance of the Jagdpanzer IV that were not directly issued to front line units were issued to the Waffenamt for testing or to various schools for troop training.

When the Allies landed in Normandy on 6 June 1944, only 62 Jagdpanzer IV were available with units in the West (31 with the Panzer-Lehr-Division, 21 with the 2.Panzer-Division and 10 with the 12.SS-Panzer-Division). The remaining 11 of the 21 issued on paper to the 12.SS-Panzer-Division in April were not released from the ordnance depot until 22 June 1944. The other divisions engaged in the West in the order that they were issued Jagdpanzer IV were the 17.SS-Pz.Gren.Div., 116.Pz.Div., 9.Pz.Div., 11.Pz.Div., 9.SS-Pz.Div., and 10.SS-Pz.Div. At the start of the Ardennes Offensive on 16 December 1944 there were still 92 Jagdpanzer IV in service with the Panzer and Panzer-Grenadier Divisions in the West.

Three German divisions engaged against the Allies in Italy were issued a total of 83 Jagdpanzer IV in April 1944. Twenty-one were issued to Fallschirm Pz.Div. 'H.G.' and 31 were issued to each Panzer Jäger Abteilung of the 3.Pz.Gren. Div. and the 15.Pz.Gren.Div. The 21 Jagdpanzer IV with the III.Abt./Pz.Rgt. 'H.G.' were the first to be committed to action on any front. All of the remaining field units that were outfitted with Jagdpanzer IV were engaged on the Eastern Front.

An indication of the survivability of the Jagdpanzer IV can be shown by the statistics from a strength report dated 30 December 1944, as shown in the table on page 9-2-27. It is readily apparent that the Jagdpanzer IV on the Eastern Front had a much higher survival factor than those in the West. This could have been due to many factors including terrain and different opposing tanks and assault guns. But, the main contributing factor should have been the higher level of combat experience in the units on the Eastern Front.

In the November 1944 edition of the 'Nachrichtenblatt der Panzertruppe', the experience of a Panzer Jäger Abteilung with Jagdpanzer IV on the Eastern Front was reported:

The Jagdpanzer IV is completely protected against armour-piercing rounds from Russian 7.62cm antitank guns and antitank rifles. In spite of numerous hits from these weapons, the Abteilung has not lost a single one of our 21 Jagdpanzer IV through enemy action.

The roles of protecting our own units from enemy tanks and supporting the infantry in their attacks were fulfilled in all cases. During attacks it was possible to hold large areas of the front without support from other units.

The hull machinegun is exceptionally effective at all target ranges, if the commander gives specific fire commands.

Attaching Jagdpanzer to units smaller than a regiment often leads to piecemeal employment and thereby unnecessary losses. Sending damaged Jagdpanzer into combat or using immobile Jagdpanzer as stationary dug-in anti- tank guns only leads to their loss. The employment of an immobile Jagdpanzer is useless. It can not turn without a running motor and robbed of mobility it will easily be captured or destroyed by the enemy. Most mechanical breakdowns can be repaired within hours or a few days, so that a fully operational Jagdpanzer can be returned to the troops ready for action.

2.6 Jagdpanzer IV Allocation

No.	Transported	Pz.Jg.Abt.	Division
31	17Mar44	Pz.Jg.Abt.130	Pz.Lehr.Div.
14	4Apr44	Pz.Jg.Abt.38	2.Pz.Div.
7	12Apr44	Pz.Jg.Abt.38	2.Pz.Div.
21	25Apr44	III./Pz.Rgt.H.G.	Pz.H.G.
10	26Apr44	SS-Pz.Jg.Abt.12	12.SS-Pz.Div.
31	29Apr44	Pz.Jg.Abt.3	3.Pz.Gr.Div.
31	20May44	Pz.Jg.Abt.33	15.Pz.Gr.Div.
21	6Jun44	Pz.Jg.Abt.49	4.Pz.Div.
21	11Jun44	Pz.Jg.Abt.53	5.Pz.Div.
21	19Jun44	Pz.Jg.Abt.92	20.Pz.Div.
11	22Jun44	SS-Pz.Jg.Abt.12	12.SS-Pz.Div.
21	24Jun44	Pz.Jg.Abt.2	12.Pz.Div.
31	30Jun44	SS-Pz.Jg.Abt.17	17.SS-Pz.Gr.Div.
21	8Jul44	SSPz.Jg.Abt.5	5.SS-Pz.Div.
21	10Jul44	Pz.Jg.Abt.228	116.Pz.Div.
10	13Jul44	Pz.Jg.Abt.41	6.Pz.Div.
10	18Jul44	Pz.Jg.Abt.19	19.Pz.Div.
21	20Jul44	Pz.Jg.Abt.50	9.Pz.Div.
31	25/26Jul44	Pz.Jg.Abt.H.G.	Pz.Div.H.G.
21	29Jul44	SS-Pz.Jg.Abt.9	9.SS-Pz.Div.
21	29Jul44	Replacements	5.Pz.Div.
21	1Aug44	Pz.Jg.Abt.61	11.Pz.Div.
11	1Aug44	Pz.Jg.Abt.19	19.Pz.Div.
21	3Aug44	Pz.Jg.Abt.543	3.Pz.Div.
11	9Aug44	Pz.Jg.Abt.41	6.Pz.Div.
28	13Aug44	Pz.Jg.Abt.87	25.Pz.Div.
6	19Aug44	Replacements	15.Pz.Gr.Div.
21	22Aug44	SS-Pz.Jg.Abt.10	10.SS-Pz.Div.
21	5Sep44	SS-Pz.Jg.Abt.3	3.SS-Pz.Div.
10	8Sep44	Replacements	9.Pz.Div.
21	11Sep44	Pz.Jg.Abt.43	8.Pz.Div.
2	20Sep44	Replacements	9.Pz.Div.
8	20Sep44	Replacements	116.Pz.Div.
21	24Sep44	Replacements	20.Pz.Div.
21	13Oct44	Pz.Jg.Abt.128	23.Pz.Div.
31	25/31Oct44	SS-Pz.Jg.Abt.4	4.SS-Pz.Gr.Div.
14	11Nov44	Replacements	11.Pz.Div.
6	20Nov44	Replacements	11.Pz.Div.

The Versuchs-leichte Panzerjäger IV from Vomag had a sloped armour hull front and forged, curved sides on the superstructure front plate. (TTM)

Right and below: A Versuchs-leichte Panzerjäger IV from Vomag with a pistol port on the right superstructure side being demonstrated to Hitler at Arys on 20 October 1943. (BSB)

For these illustrations in the operations manual D653/39, photographs of one of the Versuchs-leichte Panzerjäger IV completed by Vomag, had the muzzle brake and the second machinegun port armour cover painted over.

Above: A Versuchs-leichte Panzerjäger IV being used as a Schulungs-Fahrzeug (training vehicle) was identified as a Panzerjäger 39 by Wa Prüf 6. (BAMA)
Below: The superstructure roof plate from a Versuchs-leichte Panzerjäger IV with a ball mount for a periscope and a cap over the hole for a Nahverteidigungswaffe (close defence weapon). (TTM)

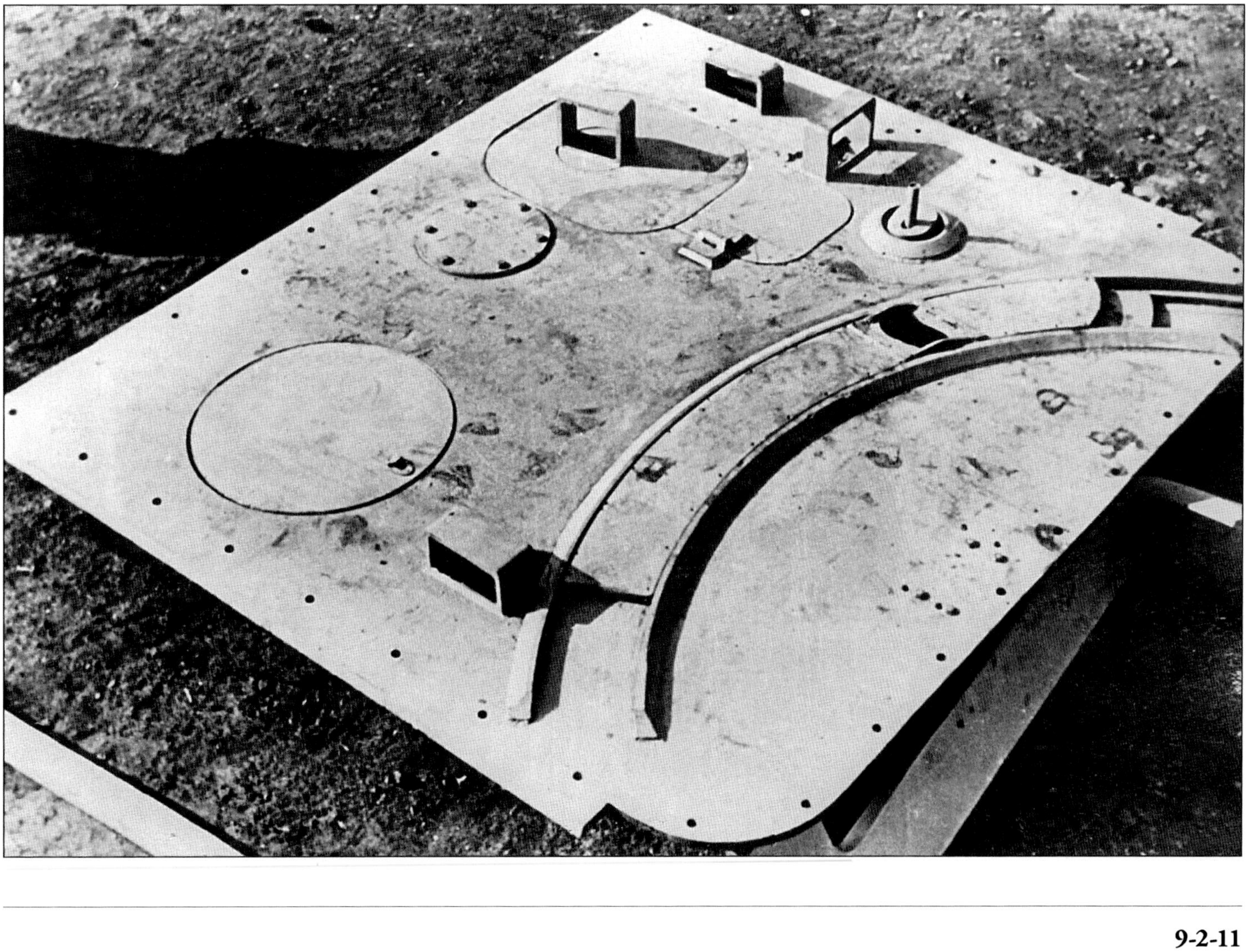

The Versuchs-leichte Panzerjäger IV, Fgst.Nr.V2 in the Golan at Saumur before being transferred to the Panzer Museum in Munster, has steel-rimmed return rollers and cast bump stops for the suspension. The ball mount base and the hole for a third periscope are closed with welded covers. (HLD)

Above: The openings in the armour guard for the driver's periscopes were machined at different angles. (TLJ)
Below: The planned forward-facing periscope and the ball mounting were welded closed when they were seen to provide little benefit. (TLJ)

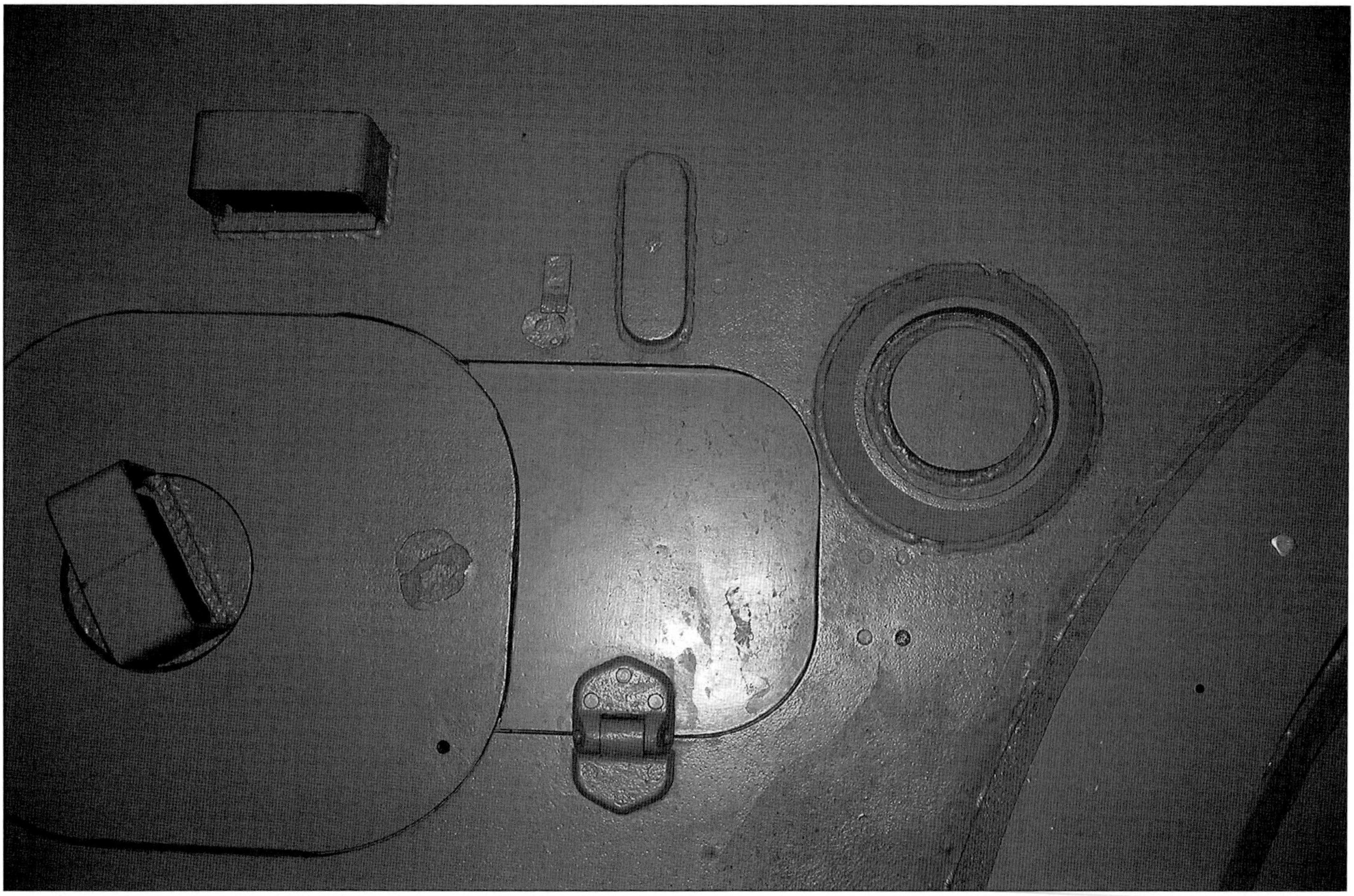

Left: The curvature of the proposed frontal armour can be clearly seen. (HLD)

Below: A Befehls-leichte Panzerjäger IV, Fgst.Nr.200020 completed in January 1944 with a Sternantenne on the left rear deck and spare track links stowed on the front. (KHM)

Pages 9-2-15 to 9-2-18: Leichte Panzerjäger IV, Fgst.Nr.320036 completed in February 1944, was retained by Wa Prüf 6 for experimental modifications including a remotely operated Rundumfeuer M.G. and a traversable as well as seven fixed periscopes around the commander's hatch on a wood superstructure deck. (BAMA)

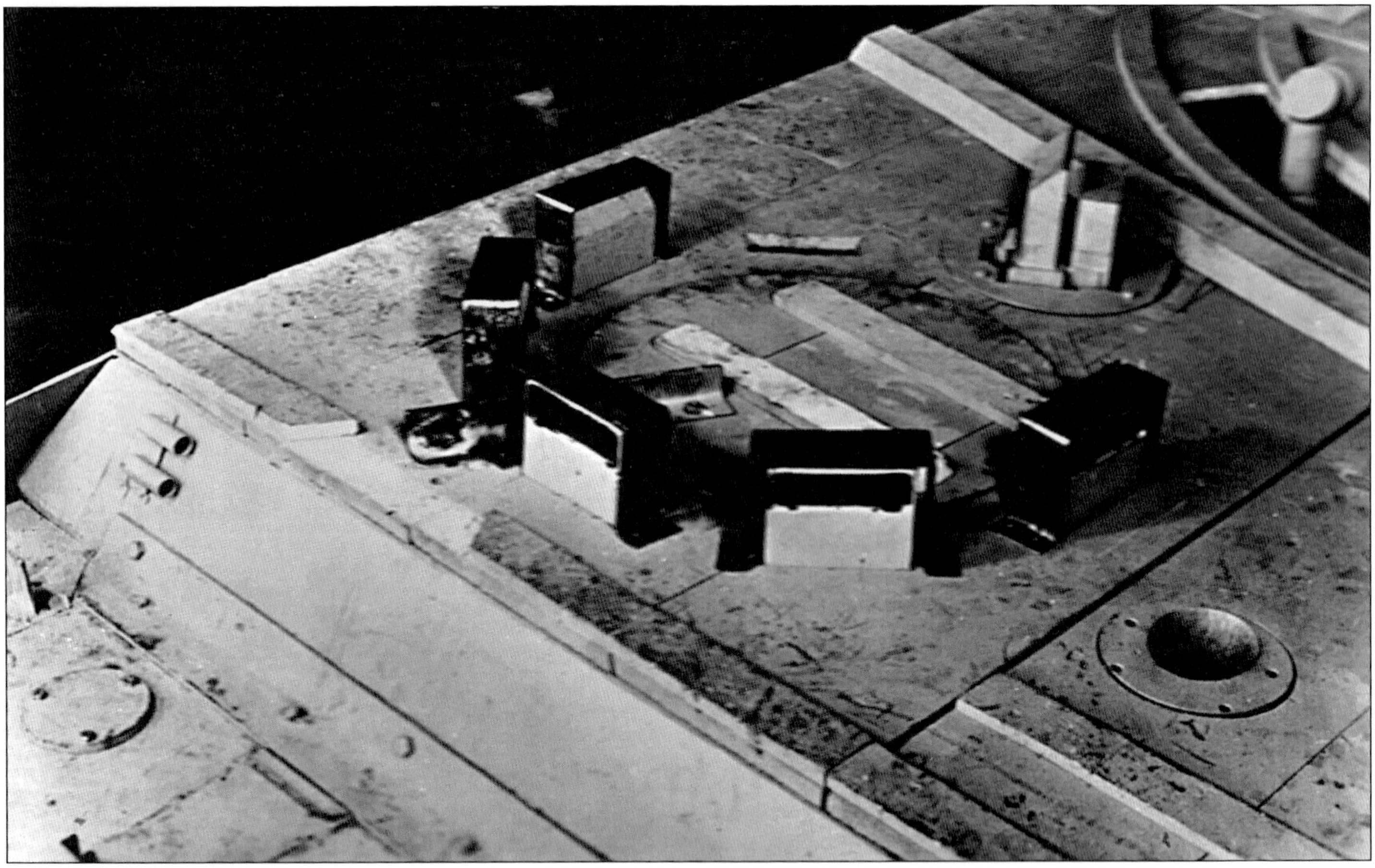

Jagdpanzer IV

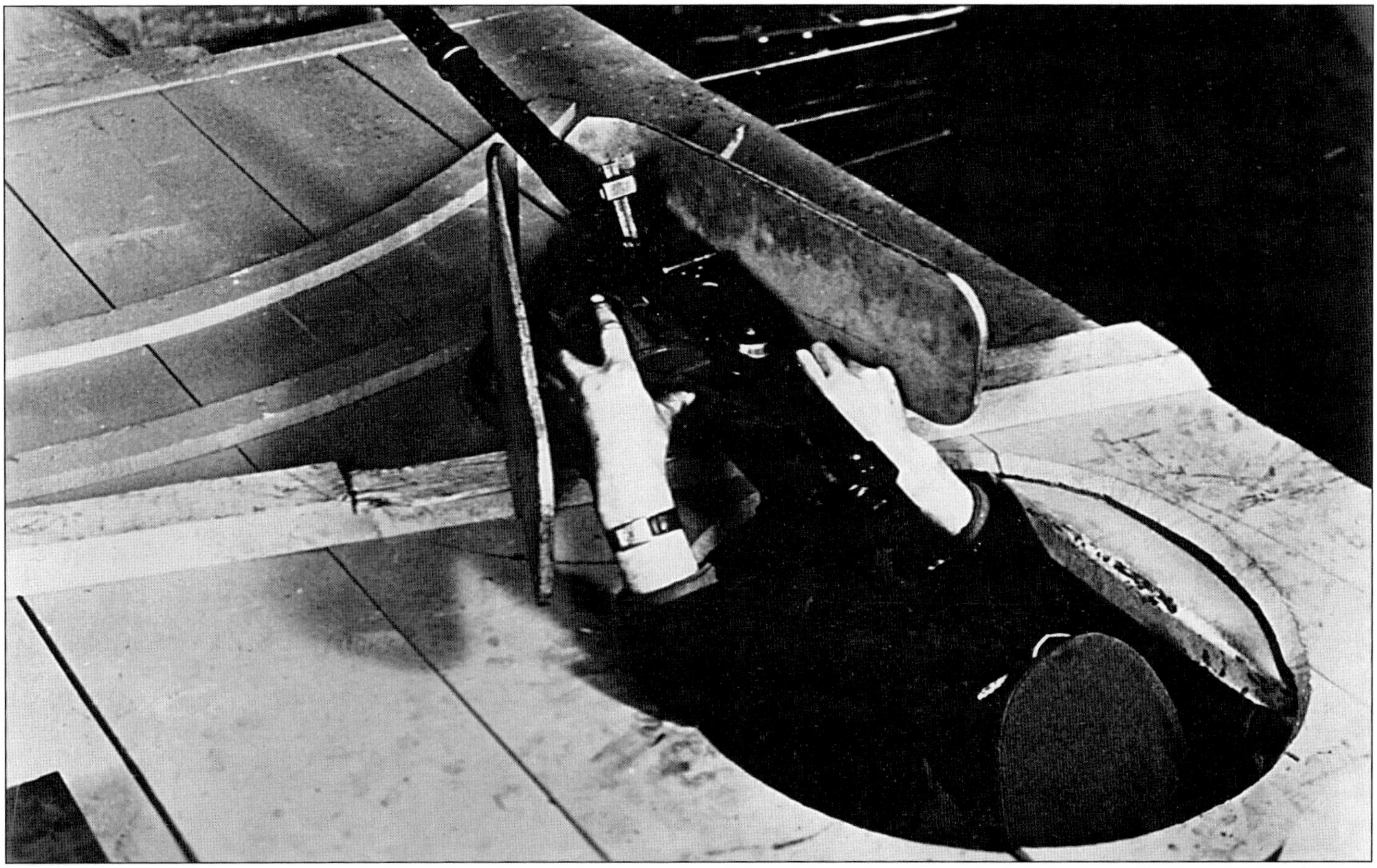

Jagdpanzer IV

Above: The experimental traversing periscope for the commander just behind the gunner's sight. (BAMA)

Left: The interior of leichte Panzerjäger IV, Fgst.Nr.320036 completed in February 1944 with an escape hatch in the belly below the gunner's seat. (BAMA)

This page and page top of 9-2-20: A leichte Panzerjäger IV completed in January/February 1944 and issued to the Ersatzheer for training with spare track links stowed on the front and spare roadwheels stowed on the upper hull rear, has an armour cover over the hole for the Nahverteidigungswaffe. (TTM)

A different leichte Panzerjäger IV with the spare track links stowed across the upper hull rear. (TTM)

Thirty-one leichte Panzerjäger IV were issued to Panzer Jäger Abteilung 130 in the Panzer-Lehr-Division and transported from the HZa on 17Mar44. This leichte Panzerjäger IV completed in March 1944 has spare track links stowed across the hull rear in a wider bracket and still does not have a Nahverteidigungswaffe (close defence weapon) on the superstructure roof. (TTM)

Thirty-one leichte Panzerjäger IV were issued to Panzer Jäger Abteilung 130 in the Panzer-Lehr-Division. This vehicle was captured in France in the Autumn of 1944. (TTM)

Above: The second machinegun port has been closed by a truncated cylindrical cover on this leichte Panzerjäger IV in Italy that was issued in April 1944. (BA 478/2188/25)

Below: As demonstrated on 8 April 1944, the muzzle brake was no longer needed on the 7.5cm Pak 39. It was retained by some units at the front but removed by others. (TA)

This page and opposite: This leichte Panzerjäger IV, Fgst. Nr.320106 completed in March 1944 was captured in Italy. It has an armour cap welded over the hole for the second machinegun, a muzzle brake on the 7.5cm Pak 39, 30mm wide strip Schürzen hangers, and the narrower bracket (the same as when the spare track links were stowed on the hull front) on the upper hull rear, and a Nahverteidigungswaffe (close defence weapon). (TTM & LA)

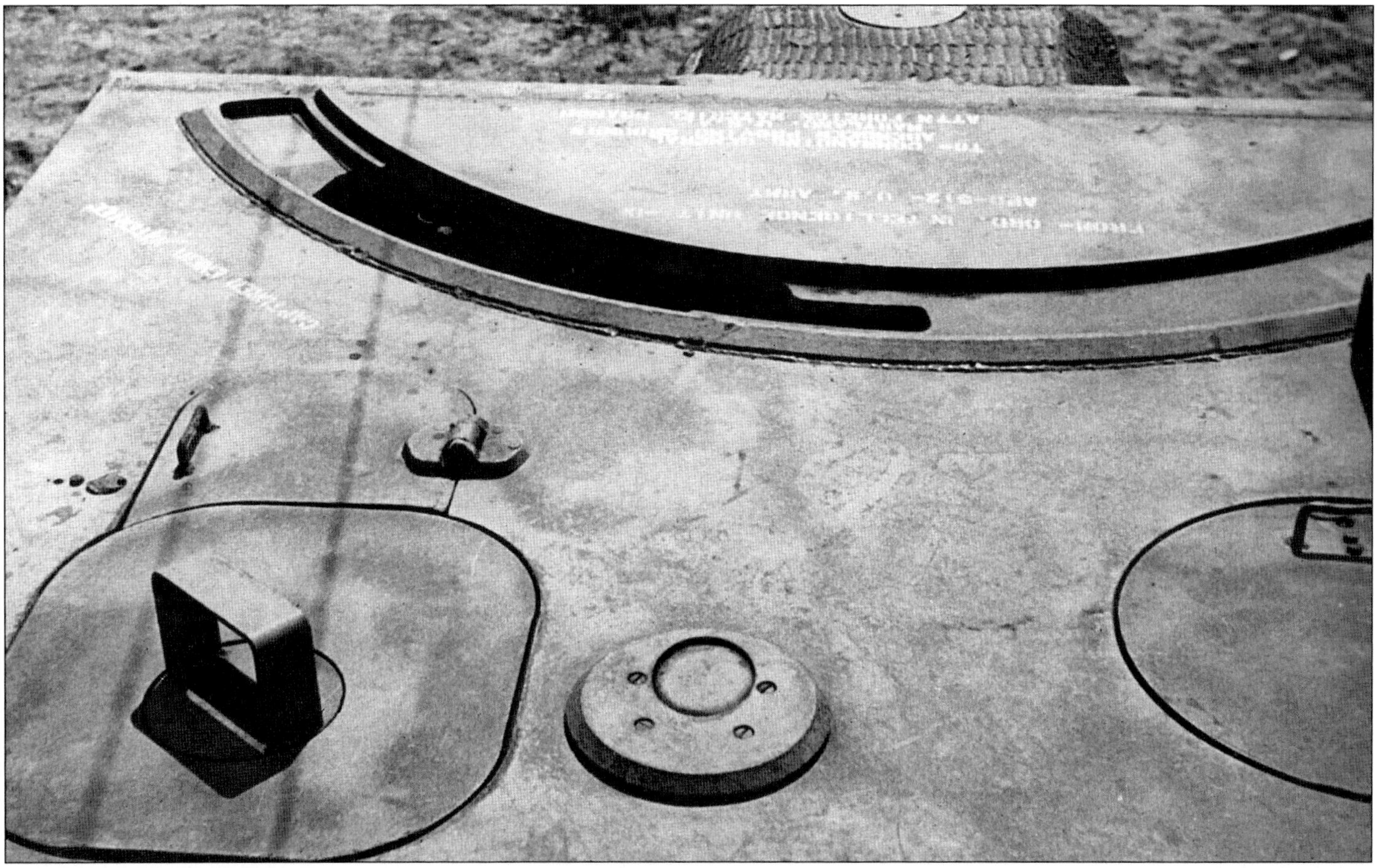

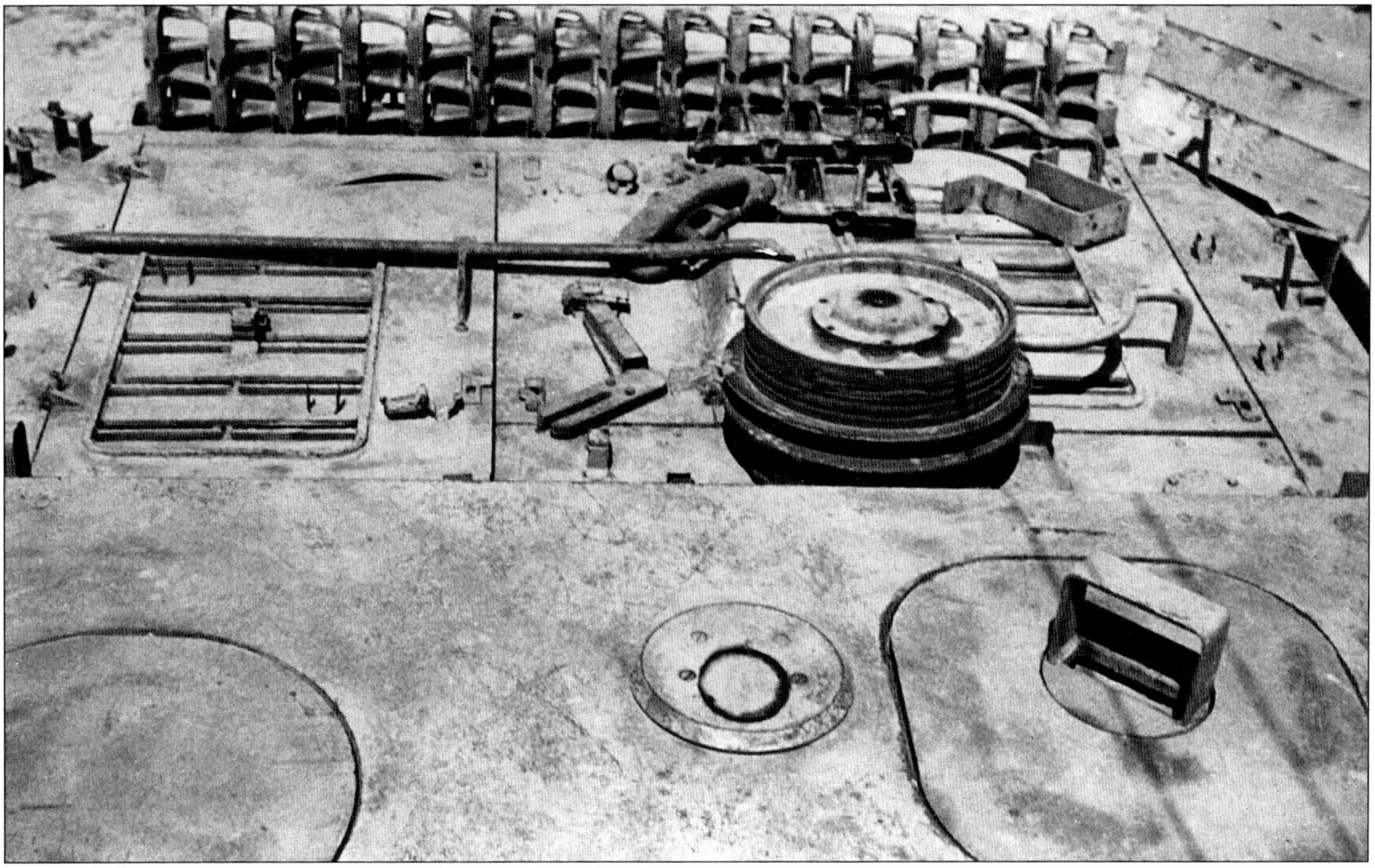

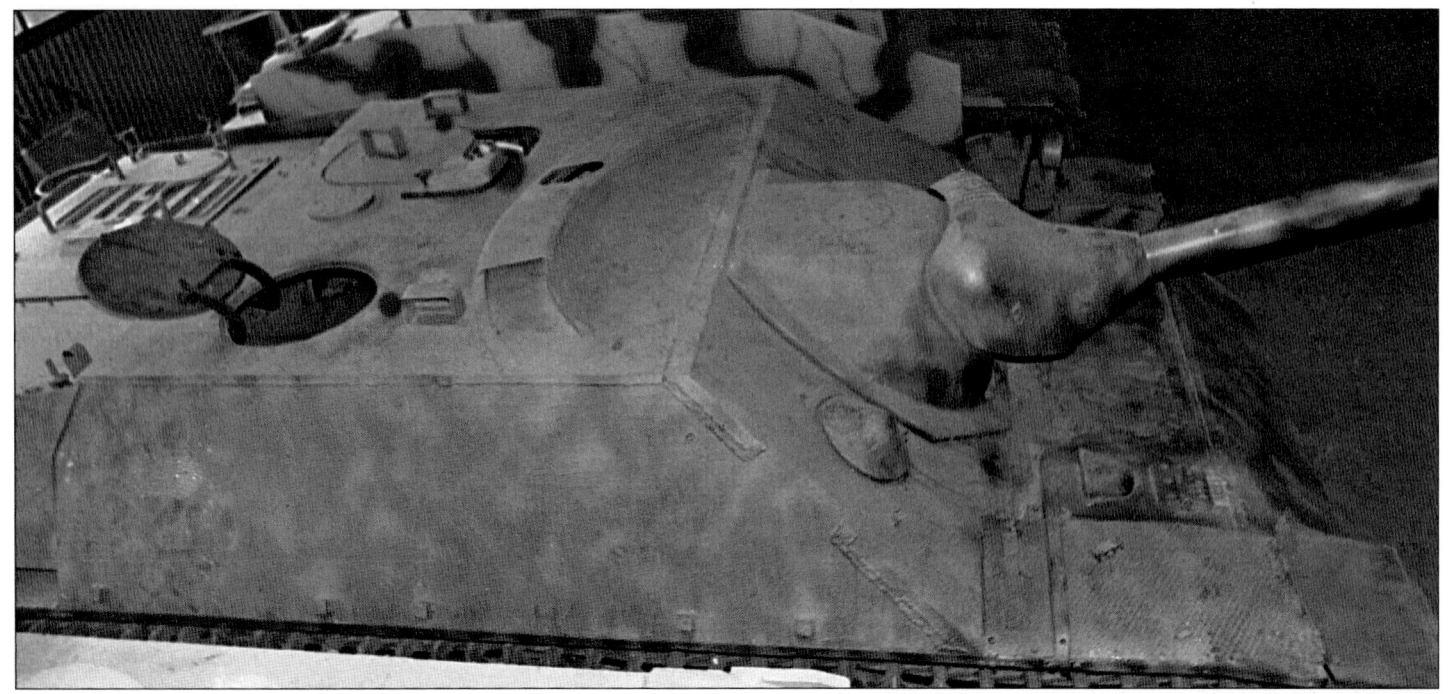

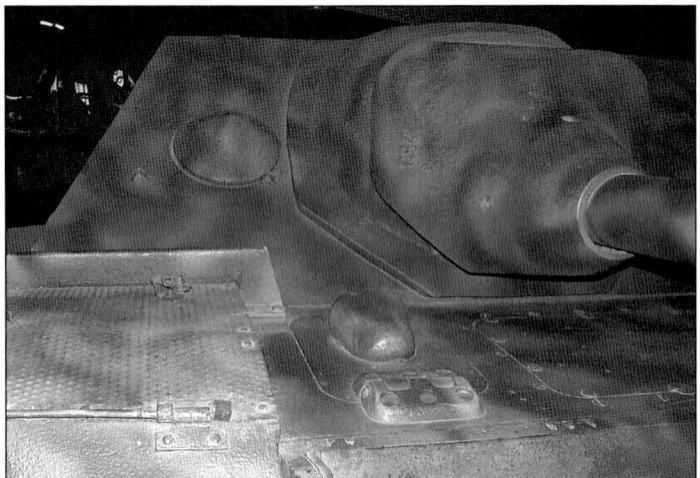

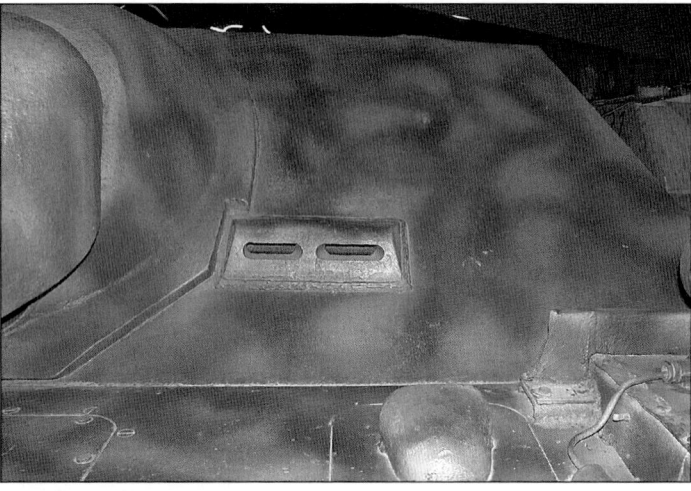

Above: This leichte Panzerjäger IV, at the Musée des Blindés in Saumur, has lost its Fahrgestell Nr. It can only be identified by its Aufbau Nr.25271 showing it was completed in May 1944. It still had 60mm thick frontal armour, did not have a second hull machinegun or a Nahverteidigungswaffe on the superstructure roof. The lower corners of the cast fixed gun mantlet were cut off to save weight. (HLD)

Left: A small number of leichte Panzerjäger IV were refurbished by the Czechs and sold to Syria. This one on display still features the wide fixed gun mantlet. (LA)

This Jagdpanzer IV (Aufbau Nr.25476) completed in July 1944 with 80mm thick frontal armour on display at the Panzer Museum, Munster, did not have a muzzle brake when issued nor did it have the spare track links stowed on the hull front or the fake Schürzen hangers. (TA)

Jagdpanzer IV Strength 30Dec44

Unit	Total	Operational	Issued to unit
Eastern Front			
H.Gr.Süd			
3.Pz.Div.	12	5	21 in Aug
6.Pz.Div.	10	5	10 in Jul, 11 in Aug
8.Pz.Div.	12	8	21 in Sep
20.Pz.Div.	19	16	21 in Sep
23.Pz.Div.	14	8	21 in Jun, 21 in Sep
3.SS-Pz.Div.	17	16	21 in Sep
5.SS-Pz.Div.	13	13	21 in Jul
H.Gr.A			
19.Pz.Div.	21	21	10 in Jul, 11 in Aug
25.Pz.Div.	21	20	28 in Aug
H.Gr.Mitte			
4.Pz.Div.	12	2	21 in Jun
H.Gr.Nord			
Pz.Div.H.G.	18	13	31 in Jul
5.Pz.Div.	25	11	21 in Jun, 21 in Jul
12.Pz.Div.	15	8	21 in Jun
Total	209	146	311

Jagdpanzer IV Strength 30Dec44

Unit	Total	Operational	Issued to unit
Western Front			
H.Gr.B			
Pz.Lehr Div.	0	0	31 in Mar
2.Pz.Div.	1	1	21 in Mar
9.Pz.Div.	6	1	21 in Jul, 10 in Sep
116.Pz.Div.	9	4	21 in Jul
9.SS-Pz.Div.	1	0	21 in Jul
12.SS-Pz.Div.	0	0	10 in Apr, 11 in Jun
3.Pz.Gr.Div.	7	0	31 in Apr
15.Pz.Gr.Div.	19	12	31 in May, 6 in Aug
H.Gr.G			
11.Pz.Div.	14	10	21 in Aug, 20 in Nov
17.SS-Pz.Gr.	2	0	31 in Jun
Total	59	28	286
Italy			
H.Gr.C			
26.Pz.Div.	8	6	
Total	8	6	

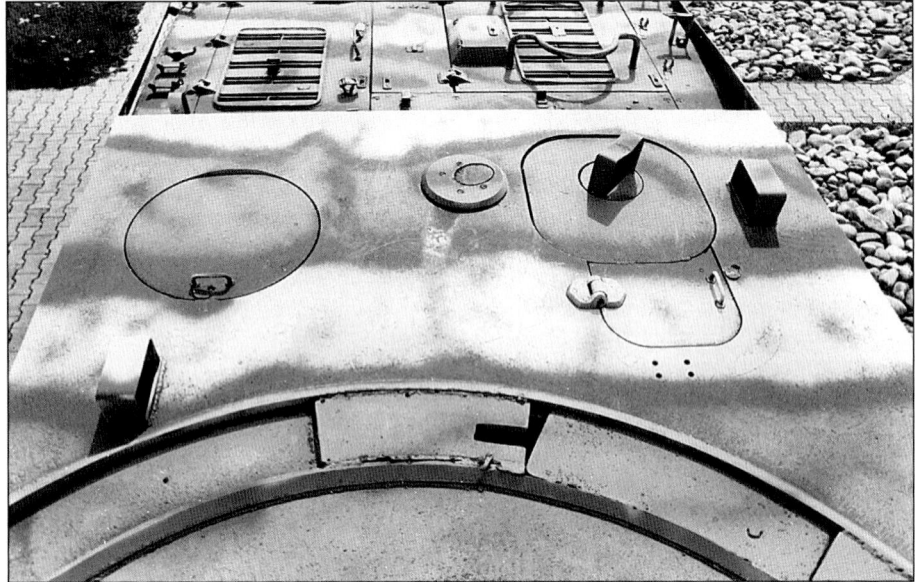

This Jagdpanzer IV (Aufbau Nr.25429) completed in June 1944 with 80mm thick frontal armour was on display at the training school in Thun, Switzerland. (HLD)

Above: Leichte Panzerjäger IV with 80mm thick frontal armour were completed starting in mid-May 1944. This one has the bent plate Schürzen hangers and the larger diameter armour guard for the machinegun. (TA)

Below: This Jagdpanzer IV, Fgst.Nr.320380, completed in June 1944 and retained by Wa Prüf 6, has 80mm thick frontal armour and later style Schürzen plates (the front and rear plates were taller). (HLD)

Above: One of the Jagdpanzer IV issued to Panzer Jäger Abteilung 19 in the 19.Panzer-Division in July 1944, has the taller Schürzen plates on the front and rear, but mounted lower in the rear. (AMC)

Below: One of the last Jagdpanzer IV completed in October/November 1944 and issued to the 11.Panzer Division, no longer had Zimmerit anti-magnetic coating or a threaded end of the barrel for a muzzle brake, and had three return rollers and tabs for securing a canvas cover over the superstructure roof. (NARA)

Jagdpanzer IV (Sd.Kfz.162) Ausf.F

Weapons & Communication

Main Armament	7.5cm Pak 39 L/48
Elevation	-8° to +14°
Traverse	12° left, 15° right
Gunsight	Sfl.Z.F.1a (5x or 8°)
Graduated to	2400m for Pzgr.39
	1400m for Pzgr.40
	3300m for Sprgr.34
	2400m for Hl.Gr.
Secondary	1x 7.92mm M.G.42
	1x 9mm M.P.40
Ammunition	79x 7.5cm Pzgr. and Sprgr.
	1200x 7.92mm for M.G.
Crew	Commander
	Gunner
	Loader
	Driver
Communication	Fu 5
	Intercom

Measurements

Length, overall	6.85m
Length w/o gun	5.90m
Width, overall	3.176m
Height, overall	1.86m
Firing Height	1.40m
Wheel Base	2.456m
Track Contact	3.515m
Combat Weight	24 metric tons
Fuel Capacity	470 litres

Armour Protection

Superstructure Front	60mm at 50° (80mm after 300th)
Superstructure Side	40mm at 30°
Superstructure Rear	30mm at 10°
Superstructure Roof	20mm at 90°
Hull Front Upper	60mm at 45° (80mm after 300th)
Hull Front Lower	50mm at 55°
Hull Side	30mm at 0°
Hull Rear	20mm at 10°
Engine Deck	10mm
Belly	2x 10mm forward, 10mm rear

Automotive Capabilities

Maximum Speed	40 km/hr
Avg. Road Speed	25 km/hr
Cross Country	15-18 km/hr
Range on Road	210 km
Cross Country	130 km
Grade	30°
Trench Crossing	2.20m
Step	0.60m
Fording Depth	1.00m
Ground Clearance	0.54m
Ground Pressure	0.86 kg/cm²
Power Ratio	11 HP/ton
Steering Ratio	1.43

Automotive Components

Engine	Maybach HL 120 TRM
	V-12 water-cooled
	12 litre petrol
	265 HP @ 2600 rpm
Transmission	Z.F. S.S.G.76
	6 forward, 1 reverse
Reverse	5.5 km/hr
1st Gear	4.5 km/hr
2nd Gear	8.6 km/hr
3rd Gear	14.5 km/hr
4th Gear	21.9 km/hr
5th Gear	31 km/hr
6th Gear	40 km/hr
Steering	Differential
Drive	Front sprocket
Roadwheels	8x2 per side
Tyres	Rubber 470/90
Suspension	Leaf springs
Track	Dry pin Kgs 61/400/120
Links per Side	99

3. CAD Drawings

Versuchs-leichte Panzerjäger IV
Fgst.Nr.V2 completed in November 1943

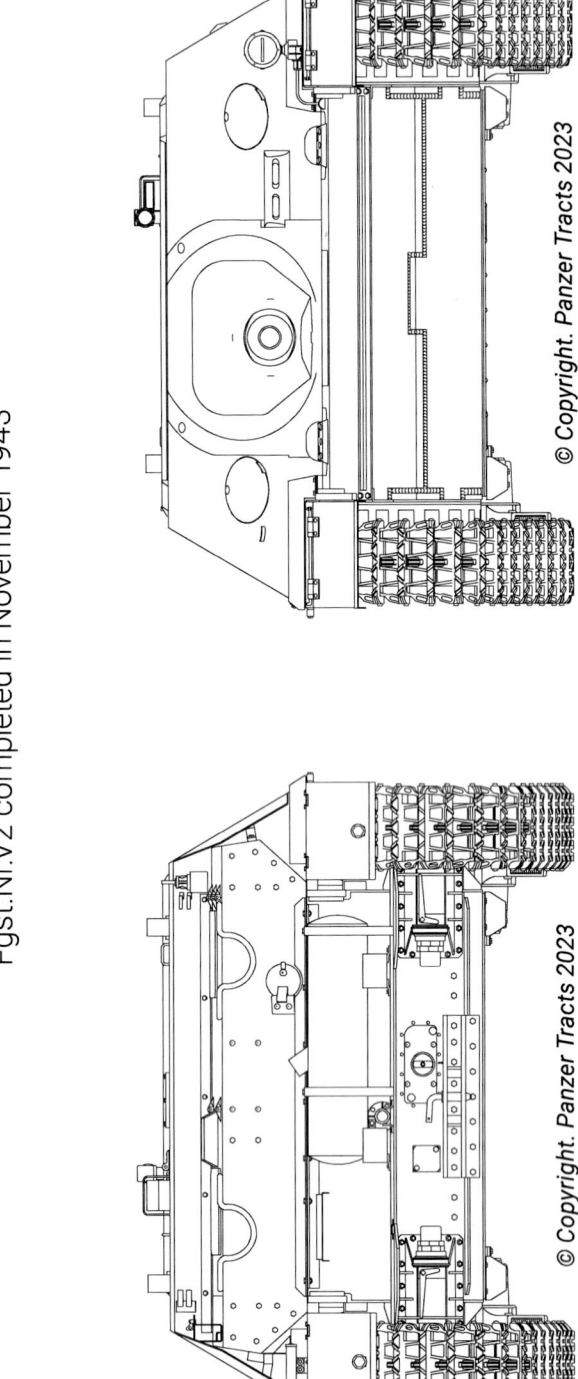
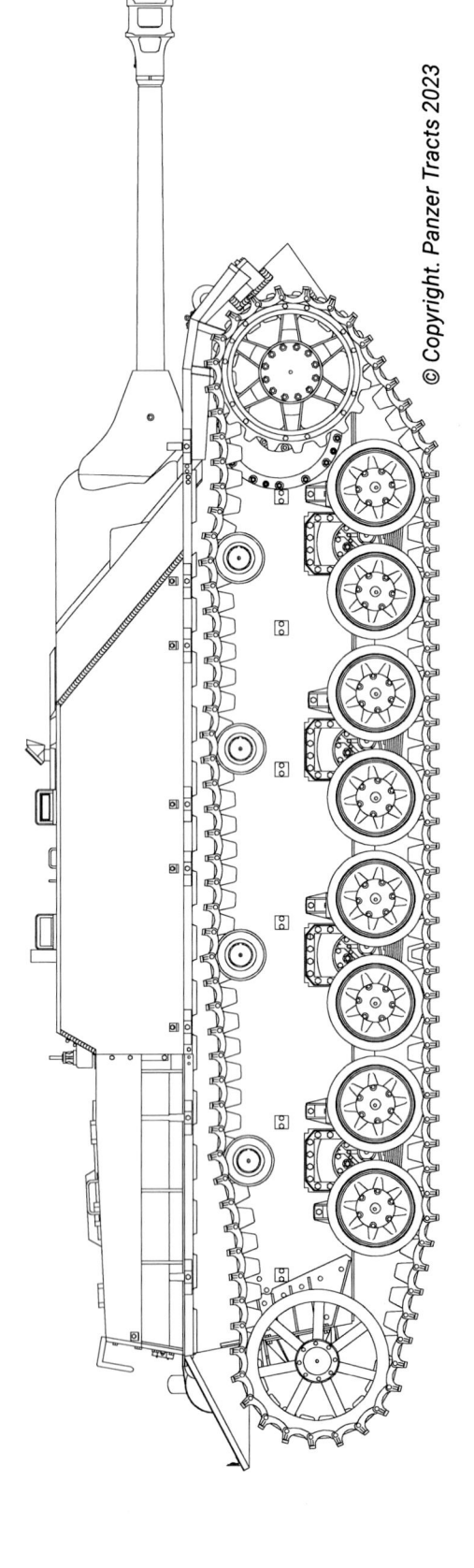

1m 2m
1:35 Scale

CAD Drawings

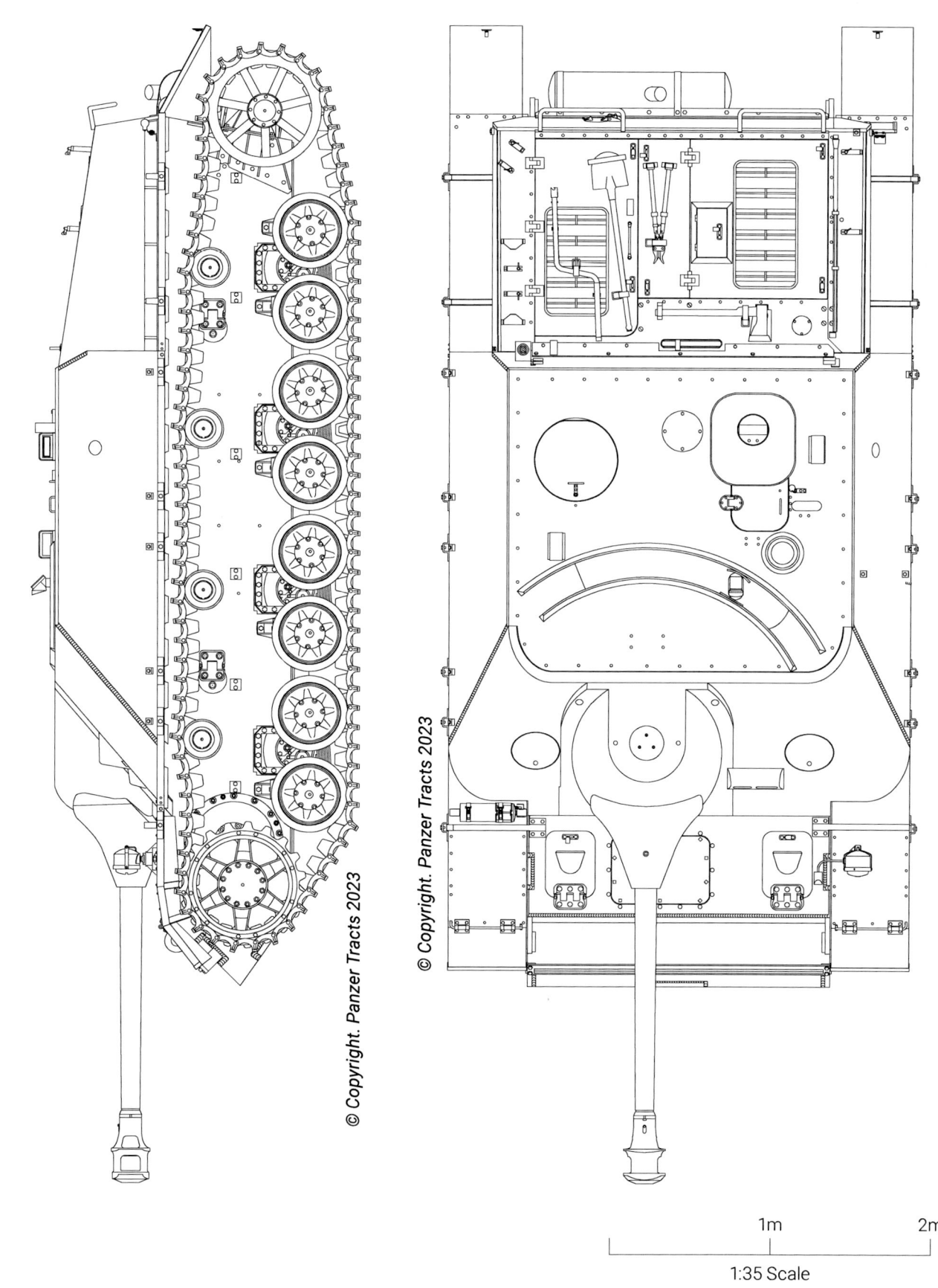

Versuchs-leichte Panzerjäger IV
Fgst.Nr.V2 completed in November 1943

© Copyright. Panzer Tracts 2023

1:35 Scale

9-2-33

CAD Drawings

Jagdpanzer IV (Sd.Kfz.162) Ausf.F
Fgst.Nr.320278 completed in May 1944

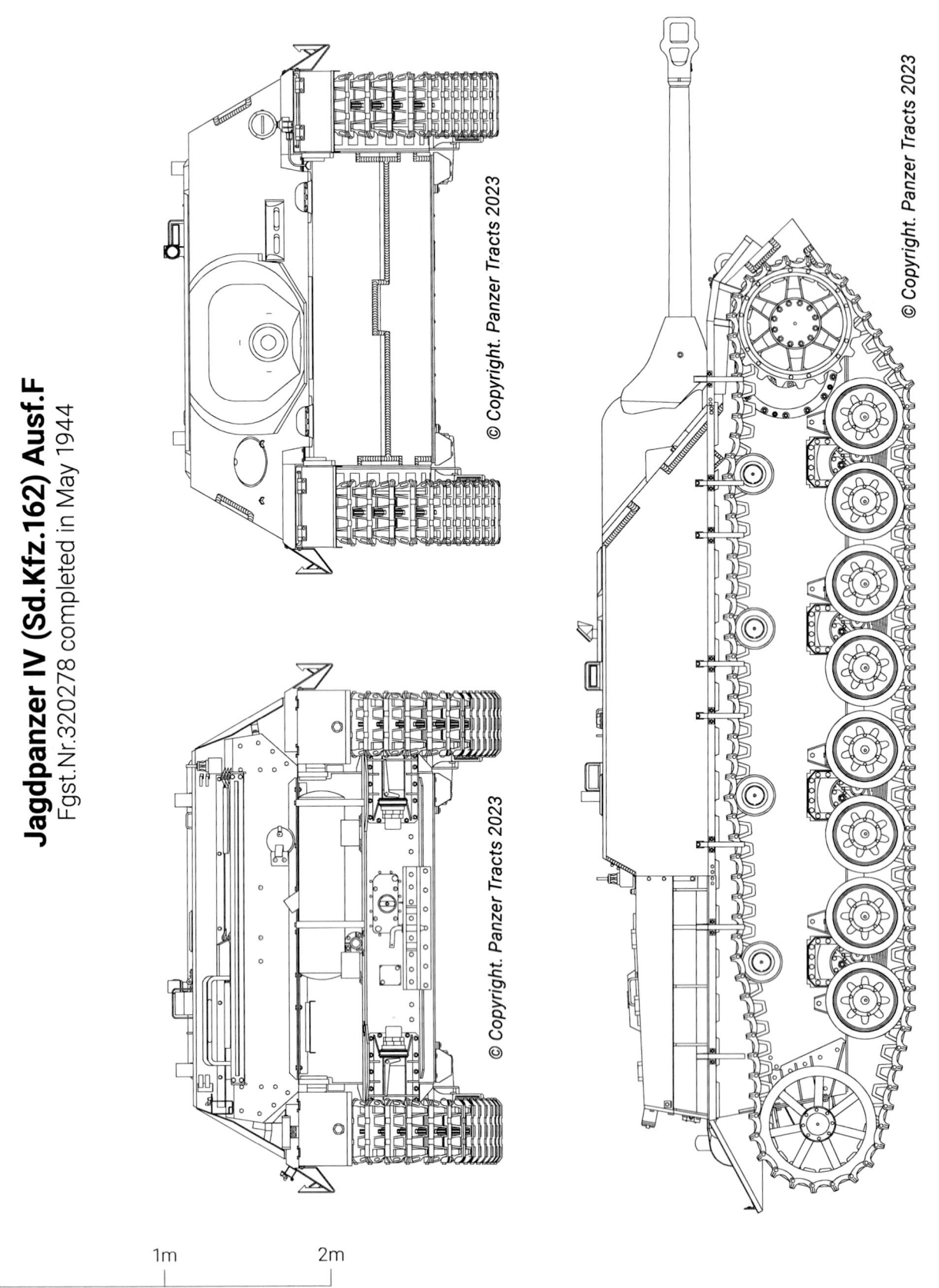

© Copyright. Panzer Tracts 2023

Features present on this Jagdpanzer IV completed in May 1944 include: a muzzle brake, 60mm thick frontal armour, a single machinegun port, a Nahverteidigungswaffe, 30mm wide strip Schürzen hangers, spare roadwheels stowed in the rear deck and spare track stowed across the upper hull rear.

1m 2m
1:35 Scale

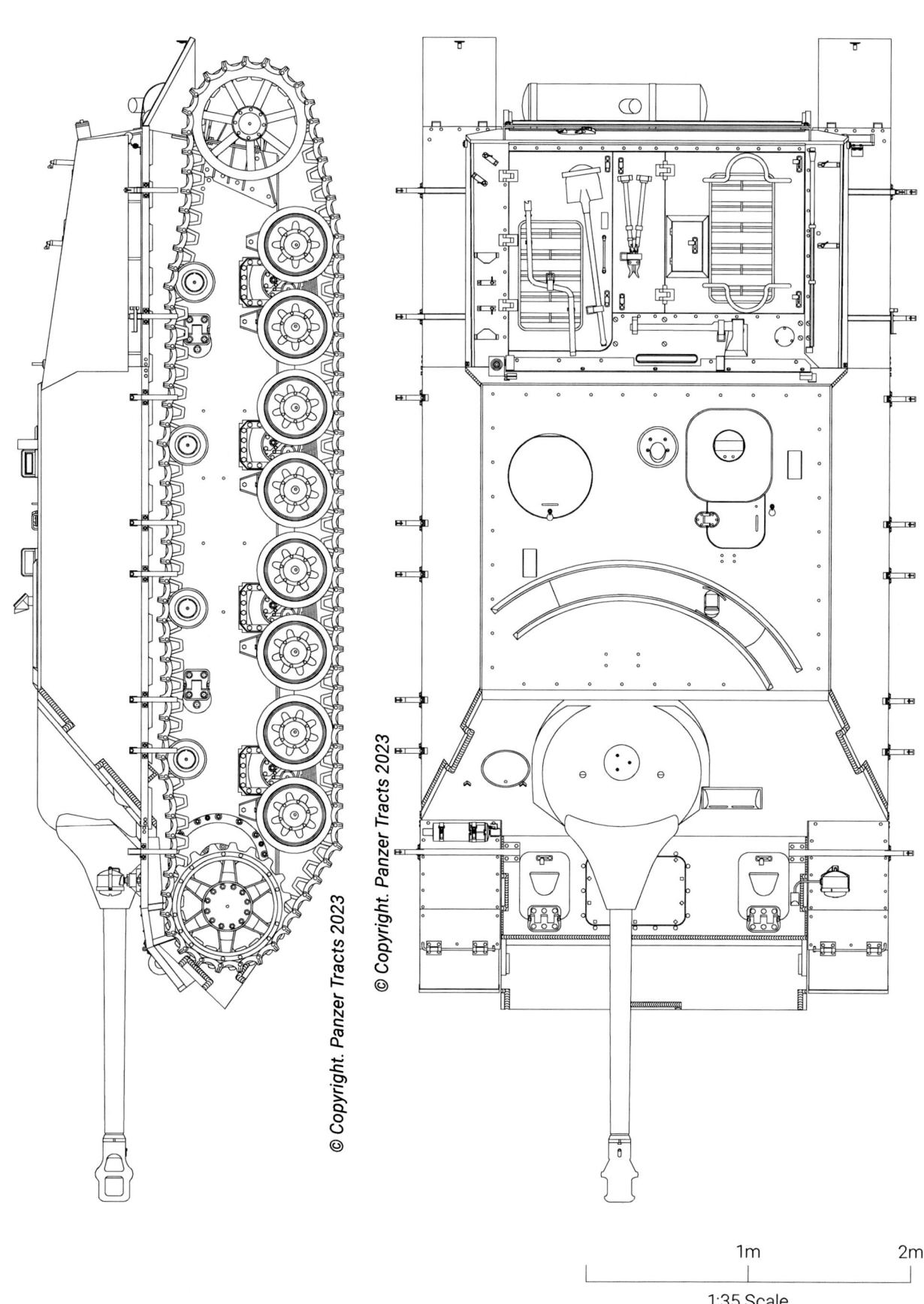

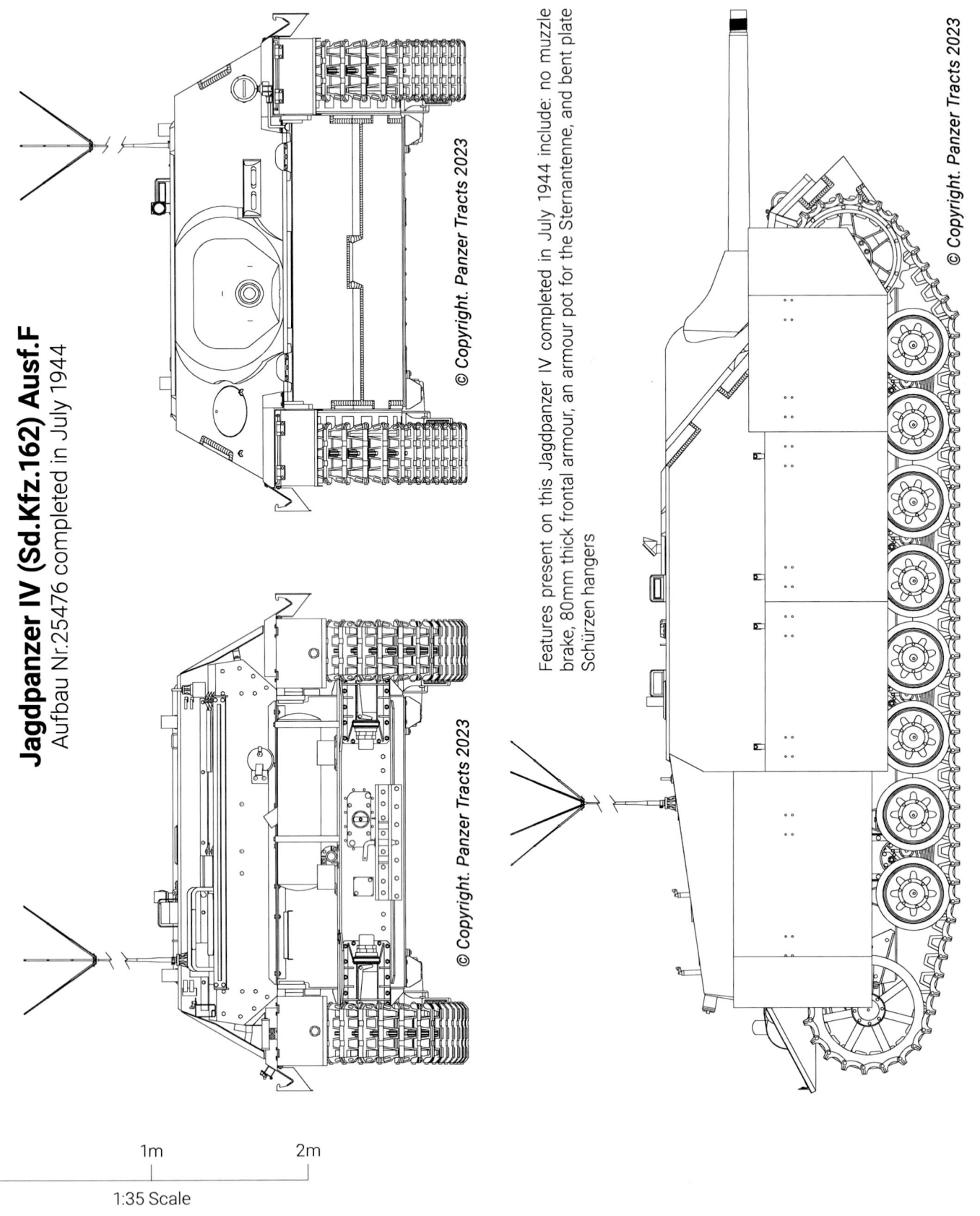

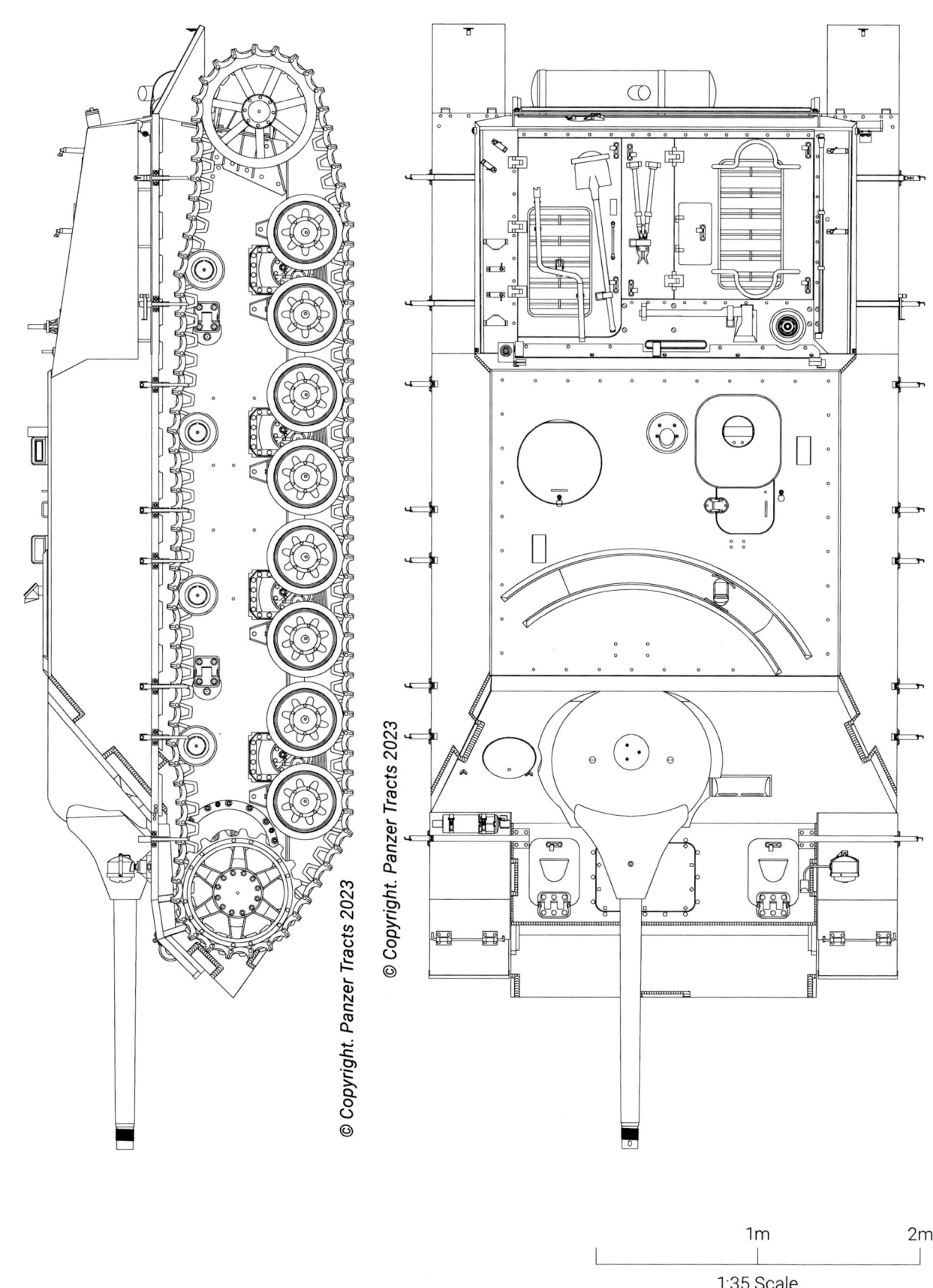

Panzer IV/70 (V)

Fgst.Nr.320864 completed in October 1944

1m
2m
1:35 Scale

© Copyright. Panzer Tracts 2023

© Copyright. Panzer Tracts 2023

© Copyright. Panzer Tracts 2023

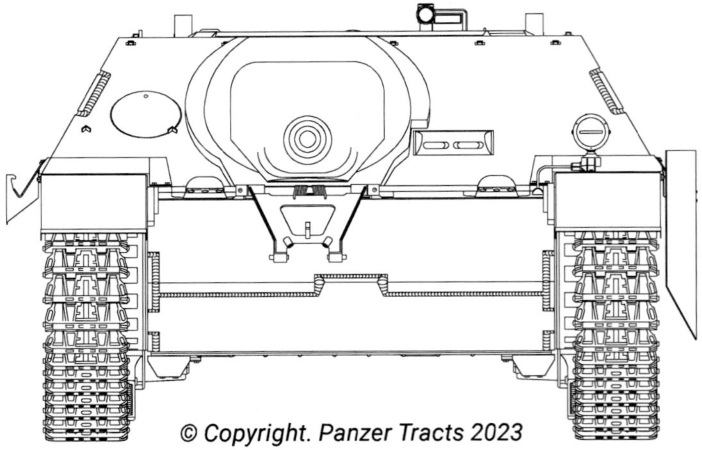

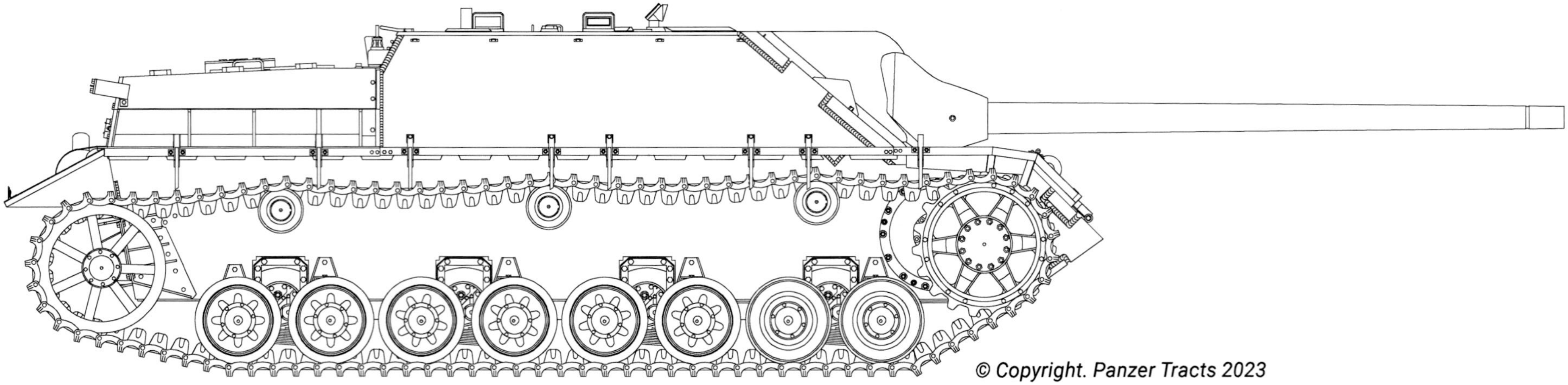

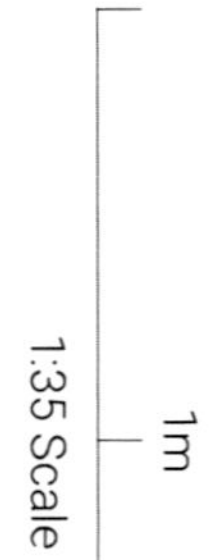

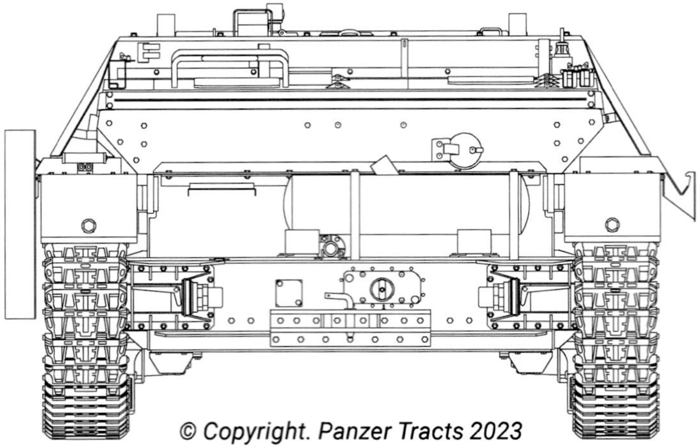

Features present on this Panzer IV/70 (V) completed in October 1944 include: an external travel lock for the 7.5cm Pak 42, angle iron segments as bump-stops for the machinegun port armour cover, tabs on the superstructure to secure a canvas cover, a shorter superstructure roof than a Jagdpanzer IV, a normal cylindrical muffler, three steel-tyred return rollers and no track-pin return plates.

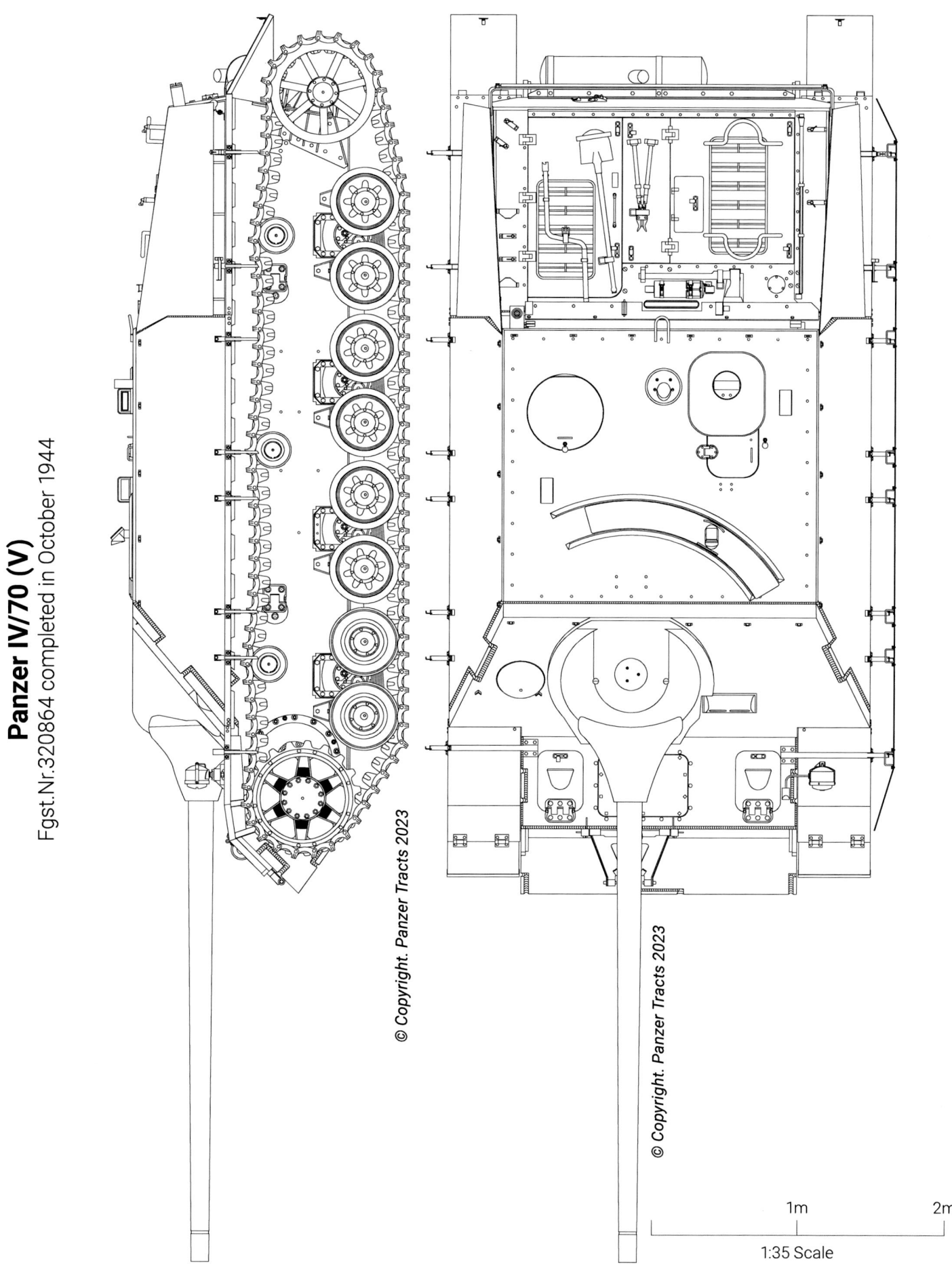

Panzer IV/70 (V)
Fgst.Nr.320864 completed in October 1944

CAD Drawings

Panzer IV/70 (V)
Fgst.Nr.329667 completed in March 1945

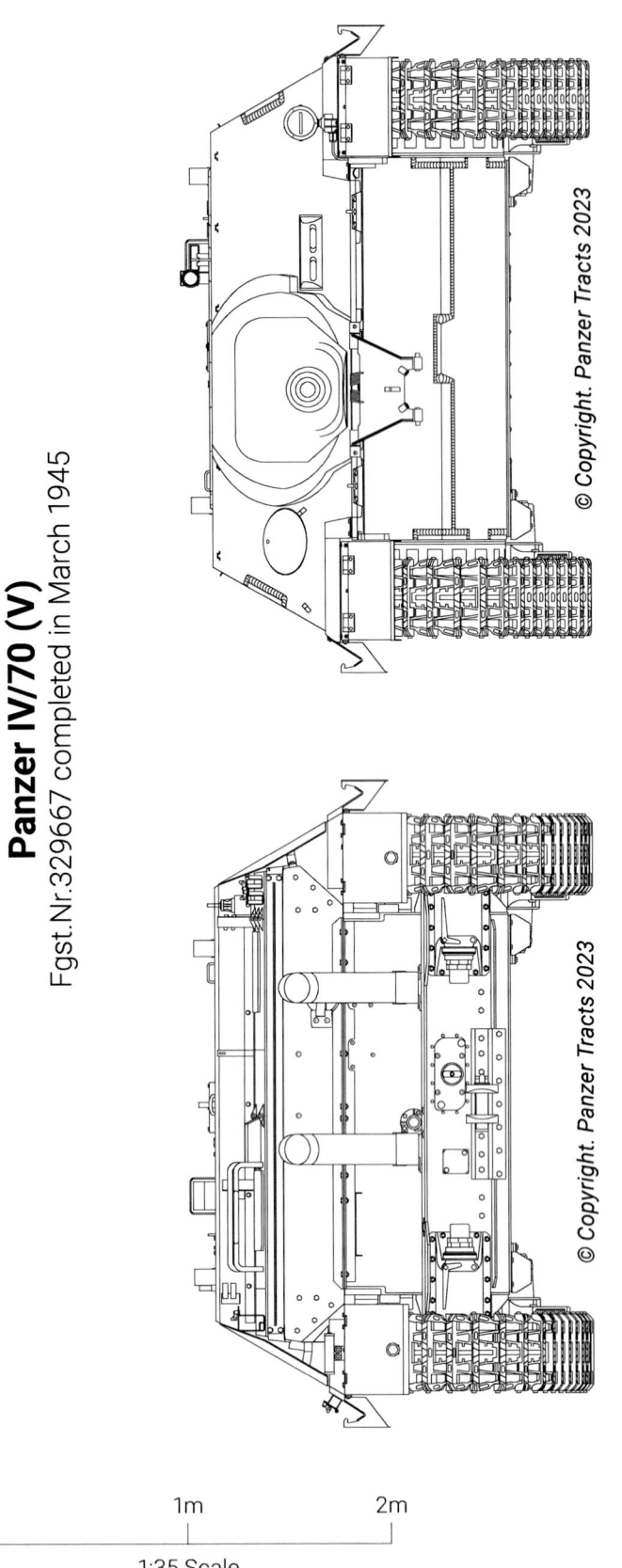

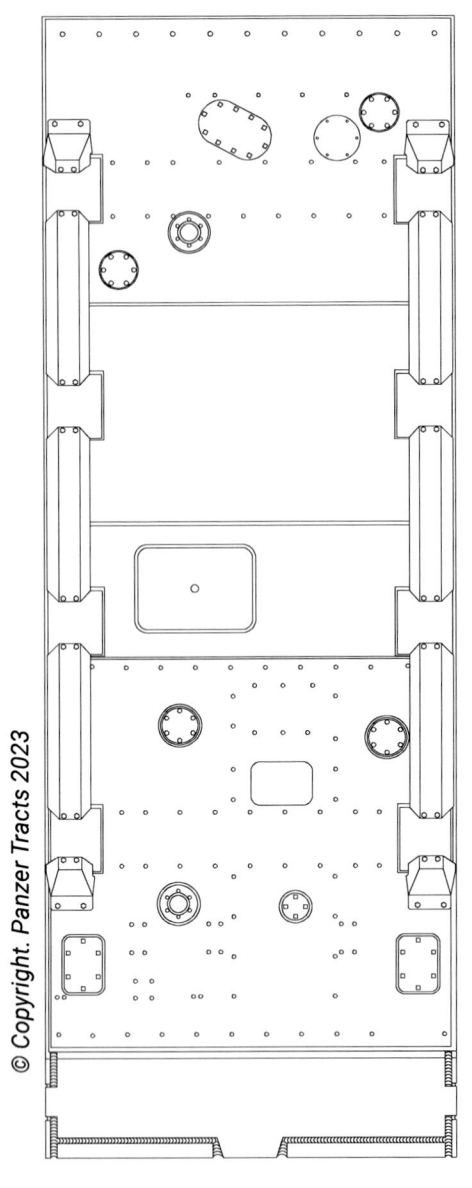

Features present on this Panzer IV/70 (V) completed in March 1944 include: handles on the brake access hatches, a segmented curved guard for the gunsight, an alignment sight for the commander, rain channels for the hatches, pegs for the EM 0.9m, Flammentöter exhausts with Krümmer and track pin return plates.

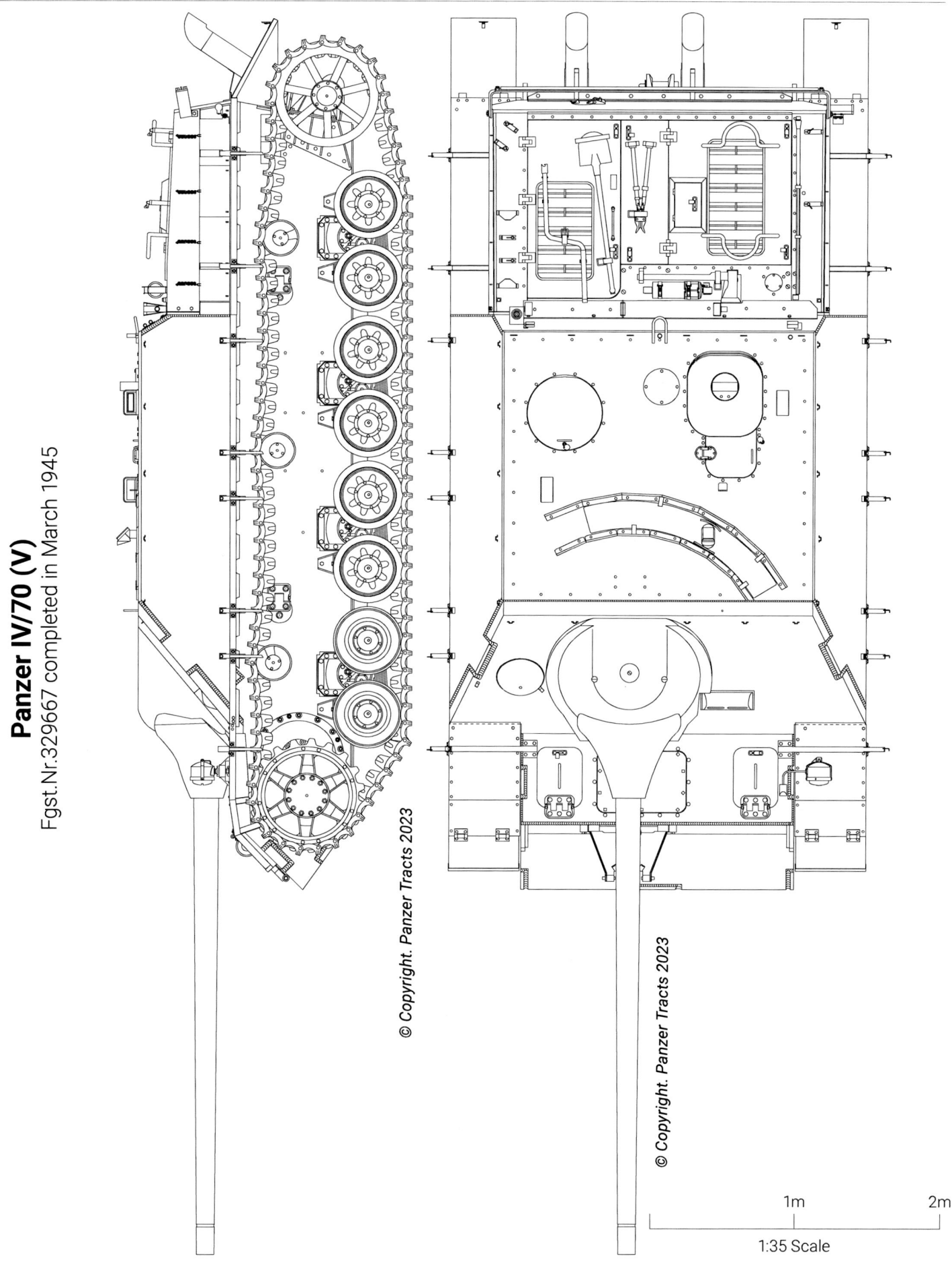

Panzer IV/70 (V) Fgst.Nr.329667 completed in March 1945

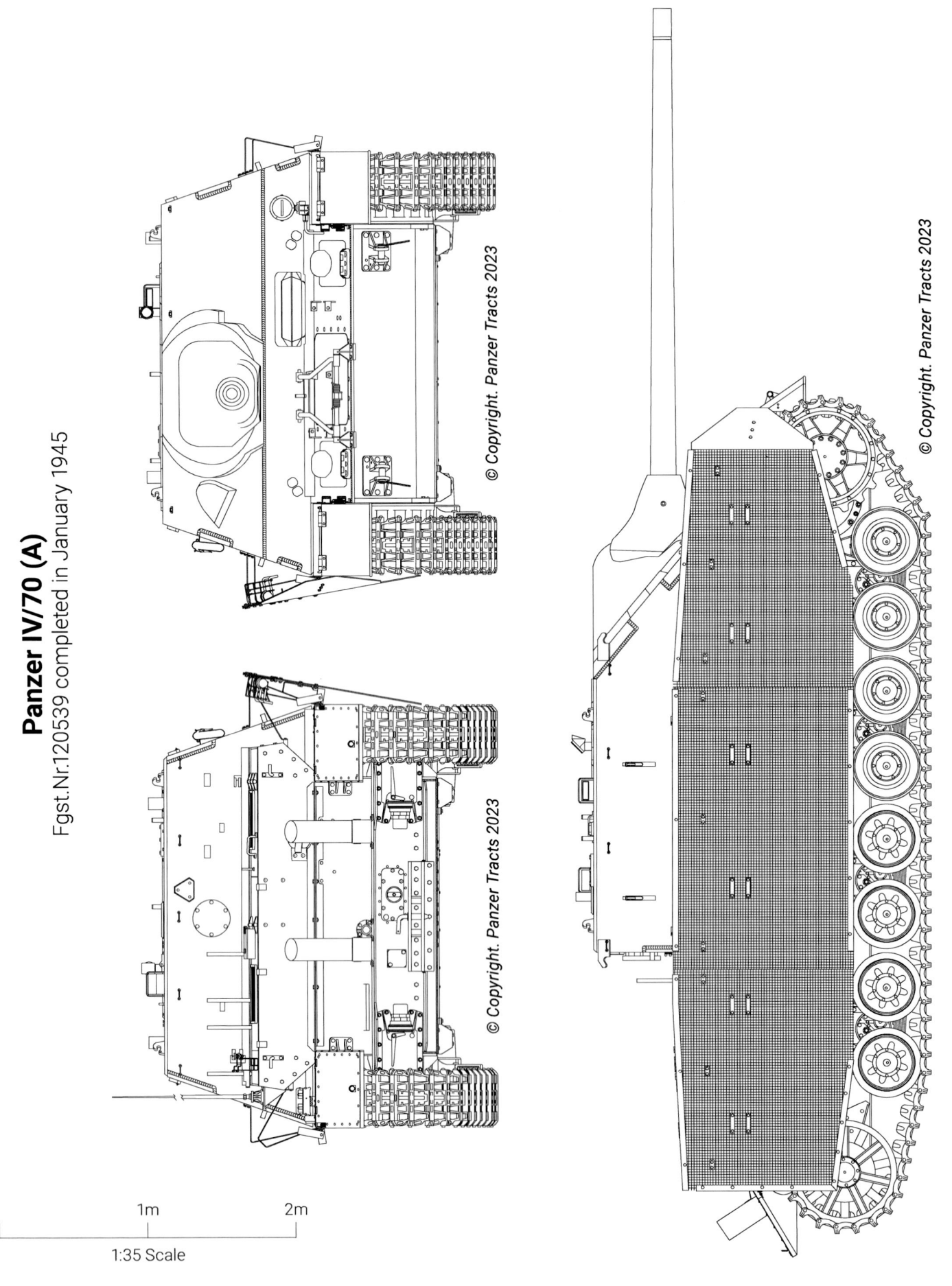

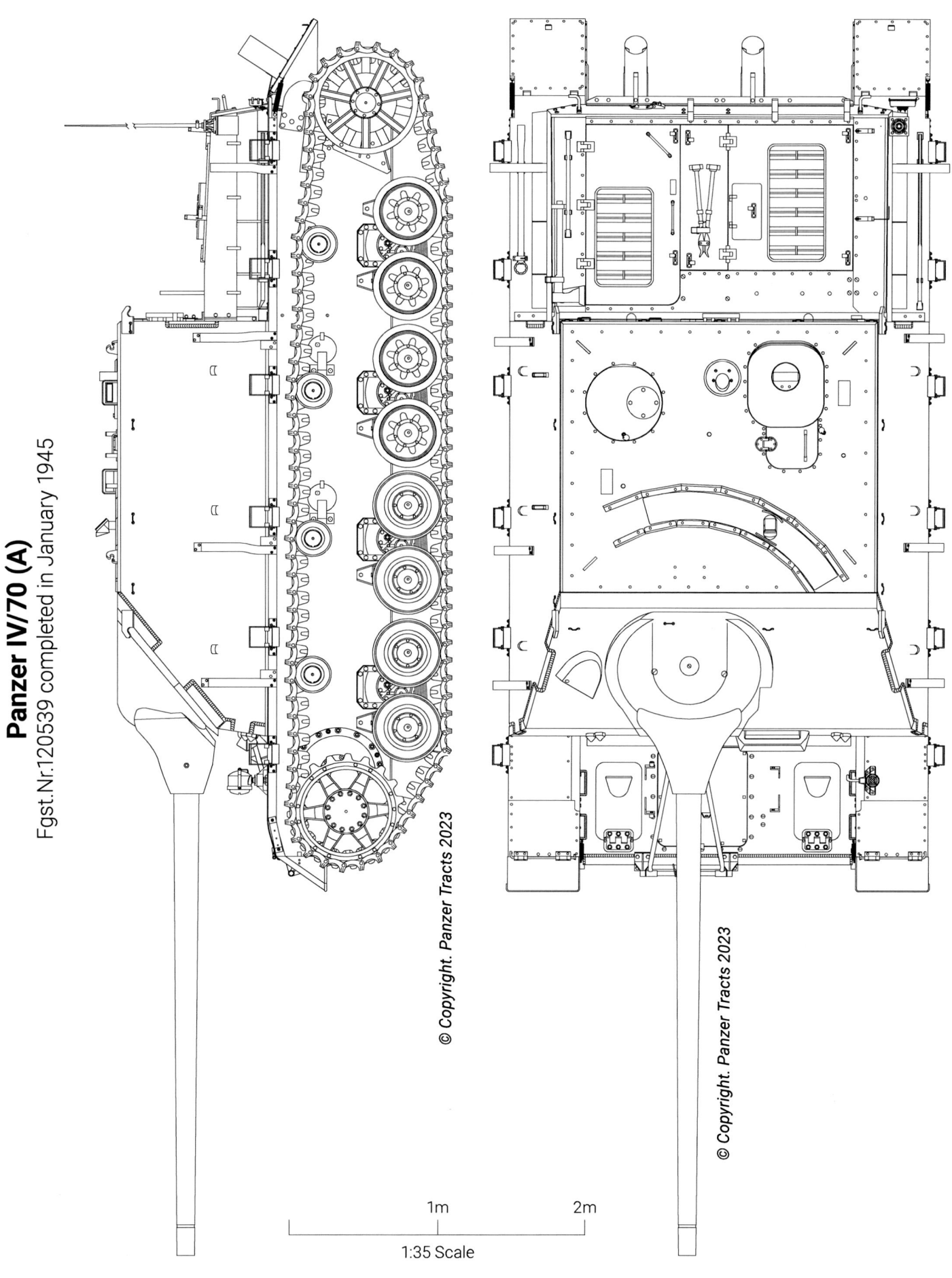

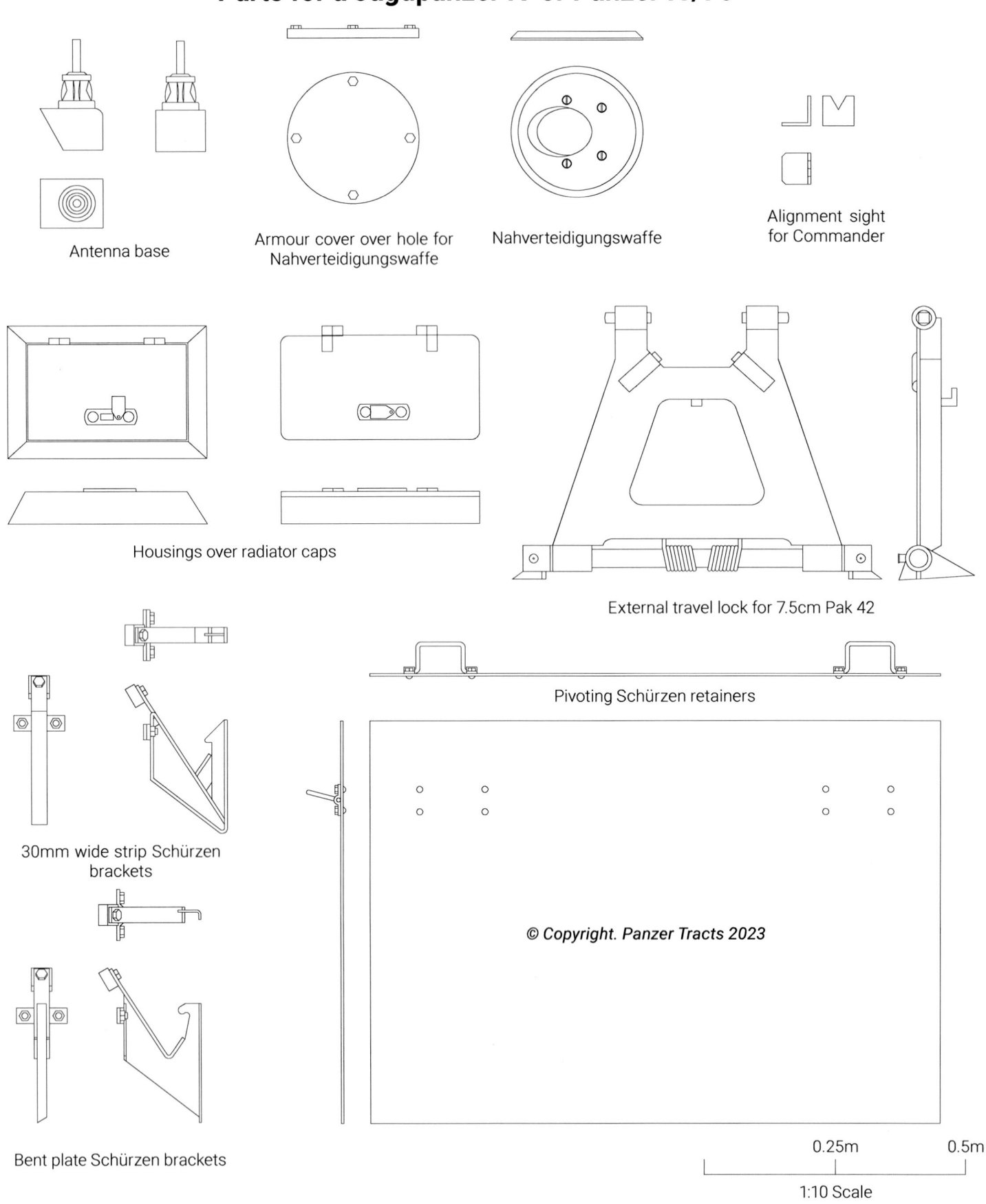

Tools and brackets for a Jagdpanzer IV or Panzer IV/70

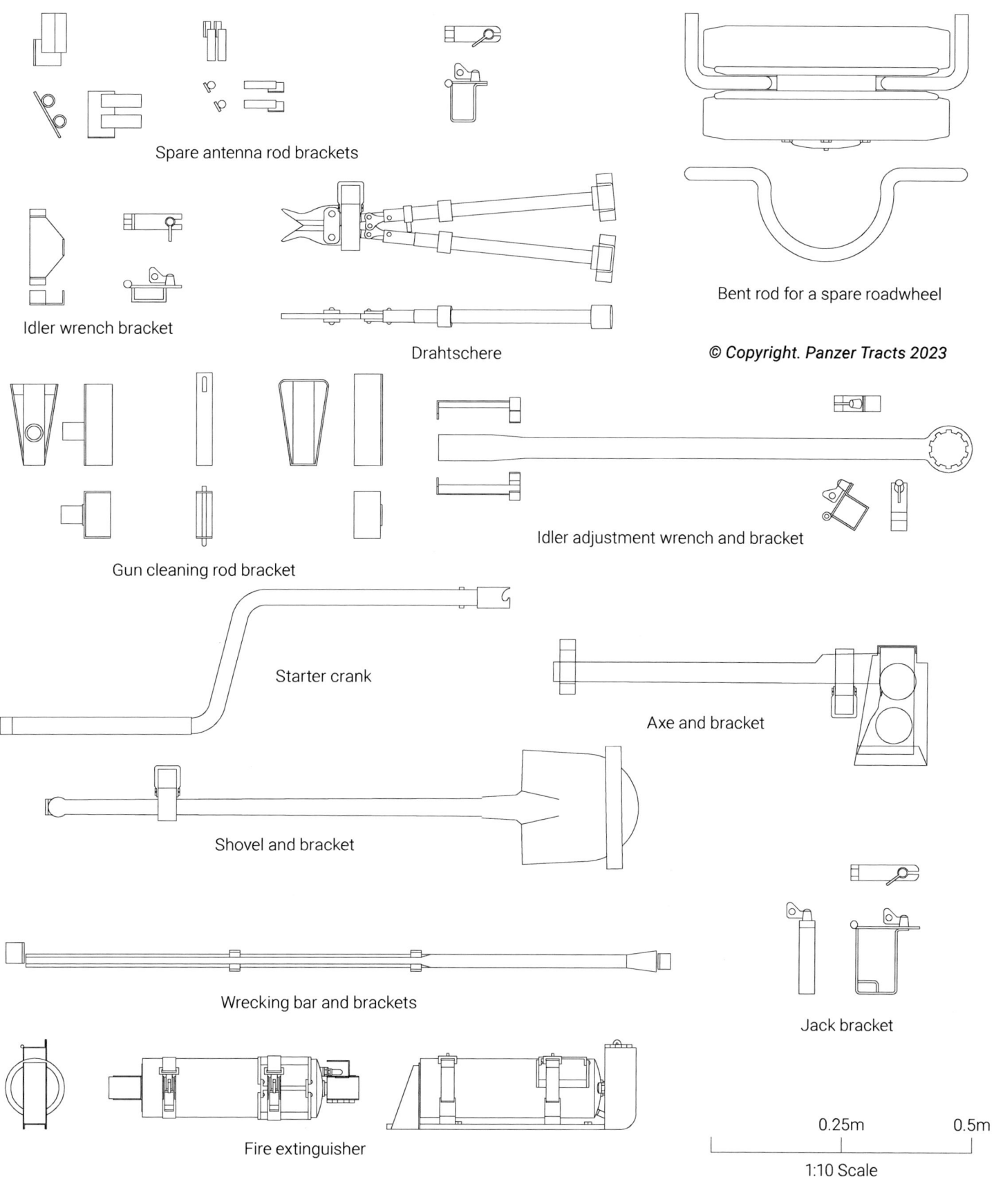

Parts for a Panzer IV/70 (A)

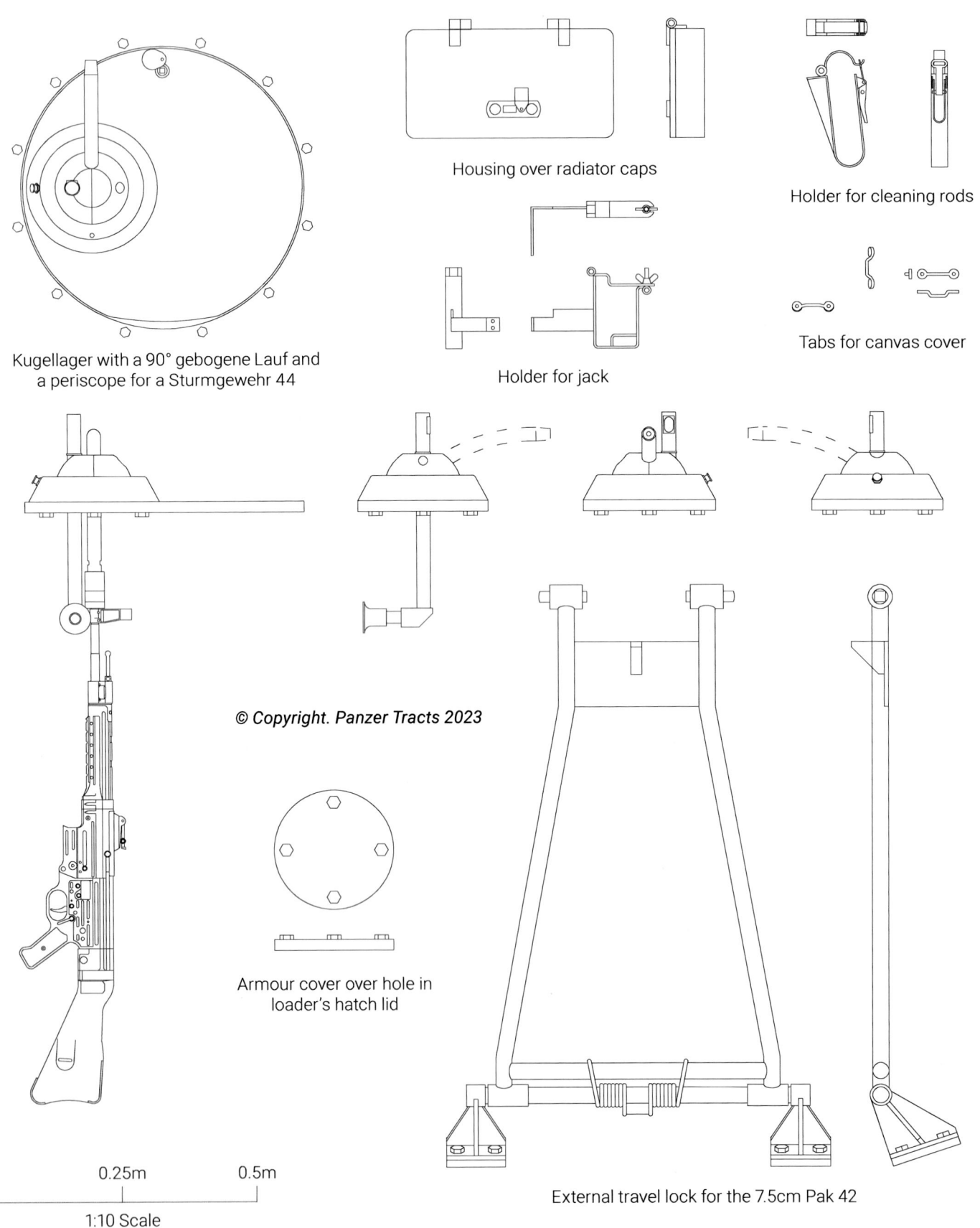

Kugellager with a 90° gebogene Lauf and a periscope for a Sturmgewehr 44

Housing over radiator caps

Holder for cleaning rods

Holder for jack

Tabs for canvas cover

Armour cover over hole in loader's hatch lid

External travel lock for the 7.5cm Pak 42

0.25m 0.5m

1:10 Scale

4. Panzer IV/70 (V) (Sd.Kfz.162)

Fgst.Nr. Serie 320651 - 321000 & 329001 - 329700

4.1 Development

The Waffenamt's initial discussions in September 1942 for a Sturmgeschütz on a Pz.Kpfw.IV chassis had included the requirement that a 7.5cm Kanone L/70 be mounted. However, this specification was not realised in the Jagdpanzer IV with its shorter 7.5cm Pak39 L/48 that started being delivered in January 1944. At a conference with Hitler held between 25 and 27 January, the 7.5cm Pak L/70 was again brought up. This longer gun was to be mounted in the Panzerjäger Vomag if it was technically feasible and if a sufficient supply of these longer guns was available.

In early April, the Waffenamt responded by mounting a converted 7.5cm Kw.K.42 L/70 in a Jagdpanzer IV (Fahrgestell Nr.320162). Hitler was shown the photographs of this vehicle at a conference on 6 April 1944 and was convinced that this Jagdpanzer with the 7.5cm Pak L/70 represented the most important recent development in armoured vehicles. A demonstration was held for Hitler on 20 April 1944 resulting in an order to increase production of the Jagdpanzer in order to reach an end goal of 800 per month. The long term program prepared by the Waffenamt on 4 May 1944 called for 2020 Jagdpanzer IV of both types to be produced from April 1944 to April 1945.

On 18 July 1944, Hitler ordered that the name for this Sturmgeschütz auf Pz.Kpfw.IV Fahrgestell would be Panzer IV lang (V). The V designating Vomag which was the firm responsible for the development of this vehicle and Vomag was also the sole assembly plant. The official names included:

- **Panzer IV lang (V) m. 7.5cm Pak 42 L/70**
  Wa J Rü Aug to Oct44
- **Panzer IV/70 (V) - Panzerwagen 604/10 (V) (m. 7.5cm Pak 42 L/70)**
  Wa J Rü Nov44 to Mar45
- **Jagdpanzer IV lang (V) (Sd.Kfz.162)**
  K.A.N. 22Jan45

The Versuchs-Panzer IV lang (V) was created by mounting a 7.5cm Pak 42 L/70 in Jagdpanzer IV Fgst.Nr.320162, which had been retained by Wa Prüf 6 for experimental modification; including mounting a remotely controlled Rundumfeuer M.G. for the loader on the superstructure roof. (HLD)

This page and opposite top: The Versuchs-Panzer IV lang (V) Fgst.Nr.320162 with the end of the barrel threaded for a muzzle brake was demonstrated to Hitler on 6 July 1944. (BSB)

Below: The Versuchs-Panzer IV lang (V) Fgst.Nr.320162 without the remotely controlled Rundumfeuer M.G. for the loader on the superstructure roof. (BAMA)

4.2 Characteristics

In order to convert the Jagdpanzer IV to mount the new gun, a series of modifications was necessary. The internal and external gun mantle were redesigned to help reduce weight while maintaining the same armour protection. An external travel lock was added to aid in keeping the sights aligned when traveling over rough terrain and to keep the end of the long gun from hitting the ground in rolling terrain. Instead of a simple exhaust fan, air blast gear was mounted to blow fumes out immediately after the gun was fired. In addition, because of the larger diameter, longer rounds fired by the 7.5cm Pak42 L/70, the ammunition storage was reduced to 57 rounds.

4.3 Production

At a meeting with Hdl. Saur on 30 May 1944, it was reported that the 7.5cm L/70 was to be installed starting with the 651st le.Pz.Jäger. Production of the 7.5cm Pjk 42 L/70 for the Panzer IV/70 (V) and (A) was reported by the Wa J Rü as 30 in July 1944, 110 in August, 107 in September, 133 in October, 210 in November, 265 in December, 258 in January 1945 and 88 in February 1945.

The first 57 Panzer IV/70 (V) were completed by Vomag and accepted by the Waffenamt in August 1944. Production continued with 41 in September and increased to 104 in October 1944, 178 in November 1944, and 180 in December 1944, reaching a peak of 185 in January 1945. Due to problems with parts supplies and power shortages, production dropped to 135 in February and 50 in March 1945. Vomag reported that production was completely halted as a result of bombing raids that hit Vomag on 19, 21 and 23 March after a total of 930 Panzer IV/70 (V) had been completed.

But this was not the end. As reported by the Abt. Org./ Führungsstaffel on 17 April 1945:

At Vomag, Plauen there are 30 Jagdpz.IV (V) Fahrgestell for which 10 superstructures are available. The II.Abt./Pz.Rgt.33 has been transferred to the Plauen area to complete and take over these Jagdpanzer.

4.3.1 Production Figures

Month	Planned	Accepted
Aug44	60	57
Sep44	90	41
Oct44	100	104
Nov44	150	178
Dec44	180	180
Jan45	200	185
Feb45	160	135
Mar45	180	50
Total	1120	930

Above and opposite: These photos of one of the first Panzer IV lang (V) in production (still unpainted and with a threaded gun barrel) were featured in an identification photo album for the Panzerjäger Ersatzheer. (TTM)

4.4 Modifications

The Jagdpanzer IV with the shorter 7.5cm Pak39 L/48 was already front heavy. This situation was further aggravated by mounting the heavier 7.5cm Pak42 L/70. On 16 May 1944, Waffen Prüf Amt 6 suggested moving the entire suspension forward by 100mm to change the centre of gravity. This suggestion was not adopted since the hull front would have to be redesigned. The mount for the front pair of roadwheels was already too close to the final drive and sprocket. Any mount that was further forward would have resulted in the front roadwheel being shredded by the drive sprocket. Another suggestion made on 10 August was to reduce the thickness of the frontal armour and to mount steel-tyred, rubber saving roadwheels as the front four on each side. Hitler on 11 August ordered that the upper front hull plate and the superstructure front plate again be reduced to a thickness of 60mm. This decision was never implemented and the frontal armour remained at 80mm throughout the entire production series. The only action that was taken to alleviate the problem of being front heavy was to mount two steel-tyred roadwheels on the frontal stations on both sides and to utilise a new lighter weight track. Both of these modifications were implemented in September 1944. In addition as part of this modification, one of the two spare roadwheels mounted on the rear deck was changed to a steel-tyred type.

Other modifications in the production series occurring in August through October 1944 were the same as the changes to the Jagdpanzer IV. Panzer IV/70 (V) Fgst. Nr.320756 completed in August/September 1944 still had Zimmerit, four return rollers per side, and all roadwheels had rubber tires; while Panzer IV/70 (V) Fgst.Nr.320864 completed in October 1944 did not have Zimmerit and had three return rollers per side and the first two roadwheels on each side were the rubber saving steel-tyred type.

Two Flammentöter (flame suppressing) exhaust mufflers replaced the large cylindrical exhaust muffler that had been mounted across the hull rear (introduced in November 1944).

Chains with rings and a spring were attached to the Kühlerabdeckklappen (cooling air intake and outlet flaps) and hooks were welded onto both upper engine compartment sides (introduced in November 1944). The Schürzenblechen (apron plates) on the sides of the engine compartment were in the way of opening and closing the flaps, so the chain was used to pull the flaps closed and the rings secured to the hooks.

Rain channels under the rims of both the loader's and

This Panzer IV Lang (V) Fgst.Nr.320756 completed in August/September 1944 was still coated with Zimmerit anti-magnetic coating, had four steel-rimmed return rollers, rubber tyred roadwheels on all eight stations on both sides, and no tabs for a canvas tarpaulin to cover the superstructure. (NARA)

commander's hatches and a segmented armour guard for the gunsight were both present on Panzer IV/70 (V) Fgst. Nr.329156 completed in December 1944 (but not on Fgst. Nr.320996 completed in November 1944).

A centred vertical Abschlepp Kupplung (tow bracket) was welded to the hull rear so that rigid towing bars could be attached (introduced in December 1944).

From a list of vision and sighting equipment dated 15 November 1944, one third of the Panzer IV lang (V) were to be outfitted with a SF 14 Z (scissors periscope) and a Entfernungsmesser 0.9m (range finder). The three pegs for mounting this EM 0.9m are present on the superstructure roof of Panzer IV/70 (V) Fgst.Nr.329667 completed in March 1945.

Five Pilze Ausf.I (sockets) were to be welded onto the superstructure roof for mounting a Behelfskran 2t (2 ton jib boom crane) to be used for lifting out component parts such as its own engine, or one from an adjacent Panzer. These were announced as a backfitted modification for the Pz.Jg. IV by In 6 on 22 June 1944. However, this modification was not implemented into the series production until March 1945. Pilze were still not present on Panzer IV/70 (V) Fgst. Nr.329667 completed in March 1945, but are present on another late production Panzer IV/70(V).

The Hutzen (air intake cowls) on the brake access hatches on the glacis were deleted and replaced with handles (present on Panzer IV/70(V) Fgst.Nr.329667 completed in March 1945). It was first discovered on 4 November 1944 that these cowls had never been necessary on the Jagdpanzer IV but had been carried over from the Pz.Kpfw.IV. A new design consisting of duct work to collect the smoke and fumes from the brakes and gears and to discharge them with the engine cooling air had been introduced with the first Jagdpanzer IV.

While every Jagdpanzer IV and most Panzer IV/70 (V) had the tubular idler wheels introduced with the Pz.Kpfw. IV Ausf.F, presumably due to a parts shortage a few Panzer IV/70 (V) produced in February or March 1945 had the heavier cast idler wheel that had been introduced on the Pz.Kpfw.IV Ausf.H.

4.5 Combat Service

The first units to receive Panzer IV/70 (V) were the newly formed Panzer-Brigades. The Panzer Abteilung of these brigades was to consist of three companies with Panthers plus a fourth Panzer Jäger Kompanie with 11 Panzer IV/70 (V).

The 105. and 106.Panzer-Brigade were each issued 11 Panzer IV/70 (V) in August 1944 and five Panzer-Brigades (107., 108., 109., 110., and Führer Grenadier) each received 11 in September. Panzer-Brigade 105 through 108 went into action in the West and Panzer-Brigade 109 and 110 saw action in the East.

In addition, in September, 10 were issued as replacements to the 116.Panzer-Division in the West and in October, 10 were issued to the 24.Panzer-Division in the East. The majority of the issue in October through December was directed at refurbishing the Panzer Jäger Abteilungen for the Ardennes offensive planned for December and for Operation Nordwind which directly followed in January 1945. Each Panzer Jäger Kompanie in the Panzer Divisions was to have 10 Panzer IV/70 (V) while each Kompanie in the Panzer Grenadier Divisions and the Heeres schwere Panzer Jäger Abteilungen was to have 14.

The following units listed in the table below are in the order that the Panzer IV/70 (V) left the Heeres Zeugamt by rail for delivery to the units.
At the start of the Ardennes Offensive there were 210 Panzer IV/70 (V) available with units on the Western Front with a further 90 reaching the troops before the end of the offensives.

In the latter half of December, attention turned again to the shape of the units struggling on the Eastern Front. The 7., 13., and 17.Panzer-Division were each issued 21 to fill their Panzer Jäger Abteilungen and the 24.Panzer-Division received 19 replacement Panzer IV/70 (V).

As the situation deteriorated, the official organisation tables were forgotten. Attempts were made to plug holes in the front lines by issuing armoured vehicles to any units that were available. Time to train the crews on a new vehicle that they had never seen before was limited to familiarity as they were driven into combat after being unloaded from a train. In this category were:

s.Pz.Jg.Abt.563	31 issued in Jan45
II.Abt./Pz.Rgt.9	26 issued in Jan45
Inf.Div.Doberitz	10 issued in Feb45
Pz.Abt.303 (Schlesien)	10 issued in Feb45
Pz.Jg.Abt.510	10 issued in Feb45
Pz.Abt.Jüterbog	10 issued in Feb45
SS Pz.Gren.Div.Nordland	10 issued in Mar45

Most of the remaining Panzer IV/70 (V) were shipped as replacements to the Eastern Front in January through March

This page: This Panzer IV/70 (V), Fgst. Nr.320864 completed by Vomag in October 1944, is shown upon arrival at the Ordnance Proving Ground Collection, Aberdeen. The application of Zimmerit anti-magnetic coating had been discontinued in September. Still visible at this time, the factory-applied original paint is the so-called ambush camouflage scheme. (USAHEC)

Panzer IV/70 (V) Fgst.Nr.320864 was displayed until recently at Aberdeen. It was repainted many times but never in the original ambush camouflage scheme. Completed by Vomag in October 1944, it features a Nahverteidigungswaffe, tabs to secure a canvas cover on the superstructure roof, two rubber-saving steel-tyred roadwheels at the front on each side, three return rollers, a rectangular housing on the rear deck over the radiator caps, and a bracket for stowing spare track links extending across the full width of the upper hull, and lighter track links with rectangular cutouts. It still had a cylindrical muffler for engine exhaust. (TLJ)

A Panzer IV/70 (V) of the 116.Pz.Div. photographed on the Brechtener Strasse, Lünen-Brambauer near Dortmund. (SAL)

1945 as shown in section 4.6 on page 9-2-63.

In a last ditch effort to shore up the defences in the West, the last 59 Panzer IV/70 (V) were shipped as replacements to various units at the front. From the first 17 from H.Za. Krugau, 5 were received by the 116.Pz.Div. and 12 by the Pz.Lehr-Div. on 26 March 1945. The next 21 from H.Za. Krugau arrived for the 15.Pz.Gr.Div. on 3 April 1945. A single Pz.IV lg.(V) from H.Za.Altengrabow arrived for Stu.G.Brig.241 on 5 April 1945. Of the last 20 from Vomag, Plauen, 6 arrived for Pz.Jg.Abt.655 and 14 for the 15.Pz. Gr.Div. on 4 April 1945.

One of the few surviving accounts relating action with the Panzer IV/70 (V) comes from the diary of a platoon leader who served in the 6.Kompanie of Panzer Regiment 9.

In late January the company received 14 Panzer IV/70 (V) at the Neuhammer troop training grounds near Sagen. The vehicles were quickly inspected, painted for winter camouflage, test driven and the guns test fired. This Jagdpanzer was the best armoured vehicle in which I sat in or fought throughout the war, especially when it came to anti-tank defence which was mostly the situation during the last months of the war.

Sent into combat in late January, after numerous engagements, the unit was sent to defend a bridgehead on the Oder river by Stettin. The platoon was attached to an infantry unit to bolster the defence and provide combat reconnaissance.

On 16 March 1945, a heavy artillery barrage began that was supported by continuous waves of 50 to 60 bombers. It was the heaviest concentration of fire that I encountered on the Eastern Front. We immediately pulled in the Scheren periscopes and machinegun and closed all the hatches. The numerous fragments flying around us couldn't affect our strong armour.

About 0900 hours, we learned that Ivan had positioned many tanks ready to attack in front of our infantry's defensive positions. After signalling the Abteilung and Regiment by radio, we learned from an infantry messenger that the rest of our Kompanie and Abteilung must already be advancing. Their progress was delayed by the ploughed up terrain caused by the heavy artillery barrage. At exactly 1100 hours, the artillery fire stopped. It was deathly still all around us. Then, from the deep holes and machinegun nests, signal flares were fired - Enemy attack! The first Russian T34/85 and SU85 rolled into the field of view of our Jagdpanzers which were in defiladed positions. Quickly flashes appeared from hits on two of the forward T34, then they were smoking. Thereafter a further five to eight enemy tanks quickly appeared beside and behind these. They burnt just as fast. So it went for most of the other enemy tanks that continued to appear in advancing tank squadrons. Every shot from our gun was now a hit. Our knowledgeable and experienced gunners, who were the oldest corporals and sergeants in the Abteilung, could hardly miss their targets. After about a 30 minute fire fight, a strong formation of T34 attempted to bypass the right flank of our position. We had fired almost all of our ammunition when behind and beside us additional guns opened fire. The rest of the Abteilung had arrived and supported our bitter defensive battle against the overwhelming Red tank formation.

Our Jagdpanzer received many hits during this difficult engagement that lasted until early afternoon. Still we remained and continued to fight until the end of the action when the last enemy tanks turned and pulled back. As the Red tank unit retired, we received a really hard hit in the final drive from a T34 that engaged us from the rear after having succeeded in getting past our flank. We lay there immobilised and had to

This page and 9-2-58: Panzer IV/70 (V) Fgst.Nr.320996, completed in November 1944 was examined by the Allies in England but is now in the US Armor and Cavalry Collection, Fort Benning, Georgia. (TTM)

wait until our friends from the Panzer Bergestaffel could tow us back to the other bank of the Oder river.

The deteriorating conditions toward the end of the war are reflected in the following report written by the commander of schwere Heeres Panzerjäger Abteilung 563 on the creation and employment of his unit during the period from 1 December 1944 to 31 January 1945.

Coming out of Kurland, the Abteilung with Stab and three Kompanien arrived in Mielau on 3 December 1944. As ordered by the Gen.Insp.d.Panzertruppen, the Abteilung was to be organised as a schwere Heeres Panzerjäger Abteilung with a Stabskompanie, one Jagdpanther Kompanie, two Jagdpanzer IV (7.5cm Kw.K.42), a Versorgungskompanie (supply), and a Panzer Werkstattzug (maintenance platoon).

On 16 January 1945, the basic training of the three Kompanien was completed. On 17 January 1945, all the combat elements of the Abteilung were employed as infantry in the Grudusk area. Casualties during this employment amounted to 55 men including specialists, commanders, gunners, and drivers.

At the beginning of operations, 150 men (armourer trainees, Panzer mechanic trainees, radio operator trainees, and crews sent to collect the Panzers) were commandeered, as well as those on leave. 35 vehicles in the Kfz. Werkstatt and 10 vehicles in the Abteilung Werkstatt were being repaired and 23 vehicles were loaned to the commandant of Mielau, which didn't return to the Abteilung.

As ordered by the Heeresgruppe, the Abteilung was to receive their weapons in Soldau. Due to a Russian breakthrough, it lost 15 special vehicles (including repair services) there. Orders were changed and the issued weapons (24 Jagdpanzer IV and 18 Jagdpanther) were sent to Allenstein. These were to be used to outfit two Kompanien each with 12 Jagdpanzer IV and one Kompanie with 9 Jagdpanther plus the attached 3.Kp./Pz.Jg. Abt.616 with 9 Jagdpanther. The shortage of crews was made good by stragglers from other units.

Raising the Abteilung began at Allenstein at 1000 hours on 20 January, and ended at 0700 hours on 21 January 1945. Due to the short time available, the Panzers could only be completely examined by the Ersatzabteilung. Range firing was not possible. Some of the drivers were sent at the last minute from the Ersatzabteilung in East Prussia. The personnel were totally exhausted from their infantry combat.

On 21 January, the Abteilung marched into action in two groups. Since then, the Abteilung was engaged in battles north of Allenstein and south and west of Guttstadt, took Liebstadt, and currently is fighting near Wormditt.

The Abteilung knocked out 58 tanks within 10 days and lost one Jagdpanther and four Jagdpanzer IV to enemy action. In addition the Abteilung lost: 8 Jagdpanther and 4 Jagdpanzer IV blown up because of fuel shortage, 1 Jagdpanther and 8 Jagdpanzer IV blown up after becoming stuck, and 3 Jagdpanther and 5 Jagdpanzer IV blown up because they needed long term repairs.

With the present personnel situation the Abteilung can immediately man 15 Jagdpanther or Jagdpanzer IV.

Panzer IV/70 (V) Fgst.Nr.320996 has solid wedge-shaped bump-stops for the cast armour guard of the machinegun port, bent wire tabs to secure a canvas cover over the superstructure roof, Flammentöter exhaust pipes, lightened track links, and track pin return plates. (TTM)

Above: This pair of Panzer IV/70 (V) were abandoned by Pz.Abt.103 at Oberpleis near Bonn on 25 March 1945. The Panzer IV/70 (V) Befehlswagen in the foreground was completed after November 1944 and has Flammentöter exhaust pipes but was outfitted with all rubber-tyred roadwheels. (NARA)

Below: A Panzer IV/70 (V) of the 11.Pz.Div., surrendered east of Bad Kötzting near Regensburg, and has most of the final features. The Hutzen (cowls) on the brake access hatches have been eliminated, a plate added to the front of the travel lock along with chains on the side of the motor compartment for raising and lowering the air intake covers. (NARA)

Above: Following the surrender of 1./s.H.Pz.Jg.Abt.655 at Oldenburg, one of their Befehlswagen is examined by Canadian troops. This Panzer IV/70 (V) completed by Vomag in February/March 1945, and is unusual as it is fitted with the cast Idler wheel. (LAC)
Below: Another Panzer IV/70 (V) of 1./s.H.Pz.Jg.Abt.655 at Oldenburg. The Panzer IV/70 (V) in the foreground has the pegs for mounting an Entfernungsmesser 0.9m rangefinder surrounding the commanders hatch. The SF14Z scissors periscope on its mount, and the Sfl.Z.F.1a gunsight has the sunshield attached. The Panzer IV/70 (V) alongside is one of the small number that have three Pilze welded to the superstructure roof for mounting a Behelfskran and Krümmer on the Flammentöter exhaust pipes. (LAC)

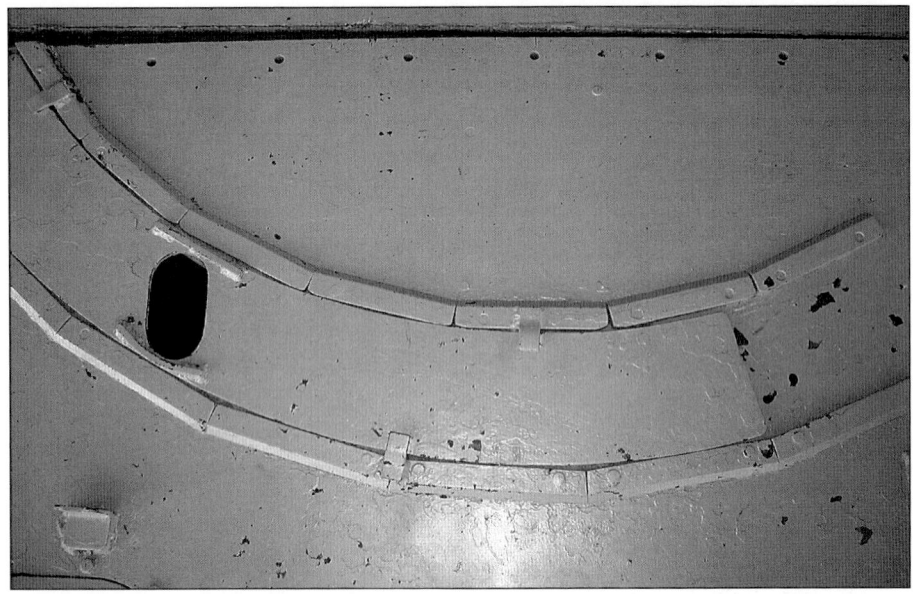

Panzer IV/70 (V) Fgst.Nr.329667, was completed by Vomag in March 1945 and is on display at the Canadian War Museum, Ottawa. Some of the late features can be seen on this example. However, the housing over the radiator caps on the motor deck has the earlier design with sloped sides indicating the reuse of old components.

Top: The cast Hutze (air intakes) on the brake access hatches have been eliminated. As a result, handles have been added to assist opening these hatches. (LA)

Middle: The guides for the sliding cover over the gunsight aperture are made from several short armour segments rather than a single machined guide. (TLJ)

Bottom: The late return rollers. (TLJ)

4.6 Panzer IV/70 (V) Allocation

No.	Transported	Pz.Jg.Abt.	Division	No.	Transported	Pz.Jg.Abt.	Division
11	1Sep44		Pz.Brig.105	2	6Jan45	s.Pz.Jg.Abt.560	
11	29Aug44		Pz.Brig.106	2	6Jan45	SS-Pz.Jg.Abt.1	1.SS-Pz.Div.
11	8Sep44		Pz.Brig.107	2	6Jan45	SS-Pz.Jg.Abt.12	12.SS-Pz.Div.
11	19Sep44		Pz.Brig.108	17	10Jan45	Reserve Mayen	
11	26Sep44		Pz.Brig.109	21	10Jan45		Pz.Gr.Div. G.D.
11	13Sep44		Pz.Brig.110	17	12Jan45	Reserve Mayen	
10	20Sep44		116.Pz.Div.	4	13Jan45	Putlos	
11	28Sep44		Fhr.Gr.Brig.	2	13Jan45	Ersatzheer	
6	6Oct44	s.Pz.Jg.Abt.560		31	15/16Jan45	Pz.Jg.Abt.563	
21	20Oct44	SS-Pz.Jg.Abt.1	1.SS-Pz.Div.	10	16Jan45		10.Pz.Gr.Div.
21	21Oct44	SS-Pz.Jg.Abt.12	12.SS-Pz.Div.	26	22Jan45	II./Pz.Rgt.9	
7	25Oct44	s.Pz.Jg.Abt.560		10	1Feb45		Doberitz
10	28Oct44		24.Pz.Div.	17	4Feb45		16.Pz.Div.
21	9Nov44	SS-Pz.Jg.Abt.9	9.SS-Pz.Div.	17	4Feb45		19.Pz.Div.
21	13Nov44	Pz.Jg.Abt.Lehr	Pz.Lehr-Div.	6	5Feb45		21.Pz.Div.
15	14Nov44	s.Pz.Jg.Abt.560		17	5Feb45		7.Pz.Div.
20	18Nov44	SS-Pz.Jg.Abt.2	2.SS-Pz.Div.	28	7Feb45		17.Pz.Div.
28	25Nov44	s.Pz.Jg.Abt.655		81	4Feb45	SS-Pz.Jg.Abt.1	1.SS-Pz.Div.
42	8Nov44		9.Pz.Div.	11	14Feb45	SS-Pz.Jg.Abt.2	2.SS-Pz.Div.
32	Dec44	s.Pz.Jg.Abt.560		12	14Feb45	SS-Pz.Jg.Abt.9	9.SS-Pz.Div.
94	Dec44	s.Pz.Jg.Abt.519		21	14Feb45	SS-Pz.Jg.Abt.12	12.SS-Pz.Div.
17	4Dec44		3.Pz.Gr.Div.	10	19Feb45		25.Pz.Div.
16	7Dec44	s.Pz.Jg.Abt.559		10	19Feb45		10.Pz.Gr.Div.
3	7Dec44	s.Pz.Jg.Abt.655		10	21Feb45	Pz.Jg.Abt.510	
3	9Dec44	SS-Pz.Jg.Abt.10	10.SS-Pz.Div.	10	21Feb45	Pz.Abt.303	
11	12Dec44		24.Pz.Div.	10	23Feb45		Pz.Div. Jüterbog
10	14Dec44		9.Pz.Div.	20	24Feb45		20.Pz.Gr.Div.
5	14Dec44		116.Pz.Div.	10	26Feb45		8.Pz.Div.
11	14Dec44	Pz.Jg.Abt.Lehr	Pz.Lehr-Div.	10	2Mar45		25.Pz.Div.
2	14Dec44	s.Pz.Jg.Abt.559		10	10Mar45		20.Pz.Div.
7	15Dec44	SS-Pz.Jg.Abt.10	10.SS-Pz.Div.	10	12Mar45		11.SS-Pz.Gr.Div.
21	19Dec44		13.Pz.Div.	5	17Mar45		116.Pz.Div.
5	21Dec44		15.Pz.Gr.Div.	12	17Mar45		Pz.Lehr-Div.
21	23Dec44		17.Pz.Div.	8	21Mar45		13.Pz.Div.
22	26Dec44		25.Pz.Gr.Div.	21	25Mar45		15.Pz.Gr.Div.
17	27Dec44		21.Pz.Div.	1	30Mar45	Stu.G.Brig.241	
21	29Dec44		7.Pz.Div.	6	1Apr45	s.Pz.Jg.Abt.655	
8	31Dec44		24.Pz.Div.	14	1Apr45		15.Pz.Gr.Div.

Panzer IV/70 (V) (Sd.Kfz.162)

Weapons & Communication

Main Armament	7.5cm Pak 42 L/70
Elevation	-6° to +15°
Traverse	12° left, 12° right
Gunsight	Sfl.Z.F.1a (5x or 8°)
Graduated to	3000m for Pzgr.39/42
	5100m for Sprgr.42
Secondary	1x 7.92mm M.G.42
	1x 9mm M.P.40
Ammunition	55-60x 7.5cm Pzgr. and Sprgr.
	1950x 7.92mm for M.G.
Crew	Commander
	Gunner
	Loader
	Driver
Communication	Fu 5
	Intercom

Measurements

Length, overall	8.50m
Length w/o gun	5.92m
Width, overall	3.17m
Height, overall	1.85m
Firing Height	1.40m
Wheel Base	2.456m
Track Contact	3.515m
Combat Weight	25.8 metric tons
Fuel Capacity	470 litres

Armour Protection

Superstructure Front	80mm at 50°
Superstructure Side	40mm at 30°
Superstructure Rear	30mm at 10°
Superstructure Roof	20mm at 90°
Hull Front Upper	80mm at 45°
Hull Front Lower	50mm at 55°
Hull Side	30mm at 0°
Hull Rear	20mm at 10°
Engine Deck	10mm
Belly	2x 10mm forward, 10mm rear

Automotive Capabilities

Maximum Speed	35 km/hr
Avg. Road Speed	25 km/hr
Cross Country	15-18 km/hr
Range on Road	210 km
Cross Country	130 km
Grade	30°
Trench Crossing	2.20m
Step	0.60m
Fording Depth	1.55m
Ground Clearance	0.40m
Ground Pressure	0.86 kg/cm²
Power Ratio	10.3 HP/ton
Steering Ratio	1.43

Automotive Components

Engine	Maybach HL 120 TRM
	V-12 water-cooled
	12 litre petrol
	265 HP @ 2600 rpm
Transmission	Z.F. S.S.G.76
	6 forward, 1 reverse
Reverse	4.8 km/hr
1st Gear	3.9 km/hr
2nd Gear	7.5 km/hr
3rd Gear	12.7 km/hr
4th Gear	19.2 km/hr
5th Gear	27.2 km/hr
6th Gear	35 km/hr
Steering	Differential
Drive	Front sprocket
Roadwheels	8x2 per side
Tyres	Rubber 470/90 + front two steel tyred
Suspension	Leaf springs
Track	Dry pin Kgs 61/400/120
Links per Side	99

5. Panzer IV/70 (A)

Fgst.Nr. Serie 120301 - 120600

5.1 Development

The Panzer IV/70 (A) is one of the least known of the German armoured fighting vehicles produced during World War II. It was outclassed by the Jagdpanther, the Jagdtiger and even its sister the Panzer IV/70 (V). In post-war publications it was given the unimpressive descriptive title of Zwischenlösung (interim solution). As can readily be seen from the few existing photographs and the one surviving example exhibited in the French armour museum at Saumur, this was a vehicle that was hastily designed without much consideration given to a clean, ballistically superior shape or low profile.

On 24 June 1944, the Heeres Waffenamt technicians at Hillersleben completed calculations which estimated the ranges at which the Pz.Kpfw.IV could penetrate the Russian T34/85 and JS122 and vice versa. To no one's surprise, they proved that the Pz.Kpfw.IV was far inferior to these enemy tanks due to both its gun and armour. The engineers at Alkett were then given an order on 26 June 1944 to create a design to mount the long barrelled 7.5cm KwK 42 L/70 on the Pz.Kpfw.IV chassis as soon as possible. A study had already been completed in September 1943 which stated five prohibitive reasons why the longer 7.5cm Kw.K. 42 could not be mounted in the turret of a Pz.Kpfw.IV. It was therefore decided that the optimum solution considering time constraints, would be to modify the superstructure of the Panzer IV/70 (V) and mount it on the existing Pz.Kpfw.IV chassis.

A second comparison study on penetration ranges, this time including American and British tanks, also came to the same conclusion that the Pz.Kpfw.IV was inferior to all the latest enemy tanks. This report dated 5 July 1944, stated that the Pz.Kpfw.IV conversion to a Sturmgeschütz mit 7.5cm Kw.K.42 L/70 would make it superior to all enemy tanks except the Josef Stalin 122.

On 18 July 1944, Hitler ordered that the name for this Sturmgeschütz auf Pz.Kpfw.IV Fahrgestell would be Panzer IV lang (A), the A designating Alkett which was the firm responsible for the development of this vehicle. The official names included:

- **Panzer IV mit 7.5cm Kw.K.L/70 auf Fgst. Panzer IV mit Aufbau Panzerjäger Vomag als Panzer IV/L (A) (lang Alkett)**
  Gen.Insp.d.Pz.Tr. 8Aug44
- **Panzer IV lang (A) m. 7.5cm Pak42 L/70**
  Wa J Rü Aug - Oct44
- **Panzer IV/70 (A) - Panzerwagen 604/9 (m. 7.5cm Pak 42 L/70)**
  Wa J Rü Nov44

5.2 Characteristics

Unlike the Versuchsfahrzeug (trial vehicle) which had the vertical lower superstructure sides, production vehicles were constructed with a simpler one piece slanted superstructure side (sloped at 20°) that continued down to the pannier over the track guards. The upper superstructure of the Panzer IV/70 (A) resembles its Vomag counterpart since the gun mount, gun shield, the slope and armour thickness of the upper superstructure front plate and the layout of the roof are the same. However, the lower front superstructure plate with the driver's visor, the superstructure sides and rear, the armoured cover for the machinegun and the ammunition racks all had to be specifically designed and manufactured for the Panzer IV/70 (A).

This superstructure had a height of 102cm compared to a height of 64cma for the Panzer IV/70 (V) superstructure. The increased height was necessitated by the fact that the fuel tanks in the Pz.Kpfw.IV hull are located on the floor under the turret. If a normal Panzer IV/70 (V) superstructure had been mounted on the Pz.Kpfw.IV hull, the 7.5cm Pak 42 gun could not have been fully elevated since the recoil of the gun and the recoil guard behind the gun would have hit the fuel tanks. Secondary reasons for increasing the height of the superstructure were: the internal gun mount interfered with maintenance on the SSG76 transmission and the muzzle of the long 7.5cm Pak 42 when mounted in the lower Panzer IV/70 (V) hit the ground in rough or undulating terrain when the gun was not in the travel lock thereby damaging the elevation gear.

Therefore, to avoid production delays which would occur if the Pz.Kpfw.IV chassis was modified, the increase in overall height of the Panzer IV/70 (A) was accepted. But this resulted in an easier target for enemy gunners due to the higher vehicle profile.

Alkett created the trial Panzer IV lang (A) by modifying a Panzer IV lang (V) superstructure to fit onto a normal Pz.Kpfw.IV chassis. (HLD)

Pages 9-2-67 to 9-2-70: The Versuchs-Panzer IV lang (A) was demonstrated to Hitler on 6 July 1944 with a trial Kugellager (ball mount) with a gebogene Lauf (bent barrel) and periscope for an M.P.44 mounted in the loader's hatch. Unlike the single plate sloped superstructure for the production series, the superstructure side for the Versuchs-Panzer IV lang (A) had vertical lower sides and sloped upper sides. (BSB)

5.3 Production

The quickly assembled, Alkett designed Versuchs-Panzer IV lang (A) was demonstrated at Berghof during Hitler's conferences on 6 through 8 July 1944. At the end of the demonstration, Hitler ordered that the following actions be taken:

The final solution is that the entire Pz.Kpfw.IV production capacity has to be converted to the Sturmgeschütz auf Einheitsfahrgestell III/IV with the long L/70 gun. Since, on the one side the present production situation won't allow this change to immediately occur, but on the other side the necessity exists to quickly deliver armoured vehicles mounting the long gun in the highest possible numbers to the front, Hitler agrees with the following transitional production adjustments:

1. In August 1944, of the planned 350 Pz.Kpfw.IV, 50 chassis are to be delivered with the Übergangsaufbau (transitional superstructure) designed by Alkett with the long L/70 gun. Because of the urgent requirements for the newly created divisions, if at all possible, the delivery of these vehicles should be pushed forward to the first half of August. b) In September 1944, through further encroachment into the Pz.Kpfw.IV production, at least 100 of these vehicles are to be delivered.

2. Immediate orders are to be issued to the armour manufacturers to convert production from Pz.Kpfw.IV hulls to Vomag Panzerjäger hulls. This will make it possible, at the latest in October 1944, to deliver at least a further 150 vehicles constructed with the long gun as the Vomag Panzerjäger, instead of the Übergangsfahrzeug Panzer IV/70 (A).

3. From October 1944 on, an increase of a further 50 conversions should be planned for each month so that the entire production capacity of 350 Pz.Kpfw.IV per month will be converted to Vomag Panzerjäger production by February 1945. As the final solution, it is self evident, that the order remains in effect that the production capacity of this factory will also be converted to production of the 'Einheitsfahrgestell'.

This 'Einheitsfahrgestell', also known as the '7.5cm Pak 42 auf leichte Panzerjäger III/IV' and as the 'Panzer IV lang E mit 7.5cm Pak 42 L/70', was scheduled to start production at Alkett and M.I.A.G. in November 1944 (phasing out the Sturmgeschütz III), at Krupp-Gruson in January 1945 (phasing out the Sturmgeschütz IV), and at Vomag and Nibelungenwerk in March 1945 (phasing out the Panzer IV/70 (V)). However, due to the necessity of maintaining delivery of as many armoured vehicles as possible, the delays always caused by starting production of a new type could not be afforded and this Panzer IV lang E was never produced.

The Waffenamt immediately implemented the intent but did not hold to the exact letter of Hitler's directive. Nibelungenwerk in Austria, the only remaining manufacturer of the Pz.Kpfw.IV at this time, received a copy of the Wasmuth Programm dated 21 July 1944 scheduling the production of 50 Panzer IV lang A in August, 100 in September, 150 in October, 200 in November, 250 in December and 300 in January. These were all ordered by the Waffenamt to be the Alkett designed Panzer IV/70 (A) with no conversion in October 1944 as ordered by Hitler to production of the Panzer IV/70 (V) but did order conversion to the Panzerjäger III/IV in March 1945. It was rare that orders relating to Panzer production that originated during Hitler's conferences were carried out exactly as stated in the minutes of the conference meetings.

A second order quickly followed with production goals set at 50 in August, 100 in September, 150 in October, 150 in November and decreasing to 100 in December and 0 in January. Conversion to the Panzerjäger III/IV instead of to the Panzer IV lang V was now ordered to begin at Nibelungenwerk in November 1944.

This was immediately followed by a third Waffenamt order in early August stating that Pz.Kpfw.IV production was to be maintained at the higher level of 250 per month and the previously planned conversion to the Panzerjäger III/IV was not to occur. This order contained the production schedule for the Panzer IV lang A of 50 in August, 100 per month from September through January, with production continuing into February 1945. As late as 30 January 1945, revised production plans for the Panzer IV/70 (A) called for 50 in January, 60 in February, 60 in March, 60 in April, 60 in May and 8 in June.

It is incredible, considering the deteriorating conditions on all fronts from the Summer of 1944, that the Waffenamt so frequently changed orders that could only have created confusion and delays in production. Switching over to the Panzer IV lang E was officially killed by a decision of the Panzersitzung on 3 October 1944 that production of the Einheitsfahrgestell III/IV was ruled out.

The Panzer IV/70 (A) which started out as an order for only 150 Übergangsfahrzeuge (transition vehicles) wound up through necessity being produced parallel to the Panzer IV/70 (V) with 277 completed by Nibelungenwerk (code 'hhv') from August 1944 through March 1945 as shown in section 5.3.1.

After determining from combat experience reports that the Panzer IV lang (A) was 'nicht fronttauglich' (not combat serviceable) on 15 January 1945 the Gen. Insp.d. Pz.Tr. proposed that normal Pz.Kpfw.IV with the 7.5cm Kw.K.40 L/48 should be produced instead.

5.3.1 Production Figures

Month	Planned	Accepted
Aug44	50	3
Sep44	100	60
Oct44	100	43
Nov44	100	25
Dec44	80	75
Jan45	50	50
Feb45	30	20
Mar45	0	1
Total	**510**	**277**

5.4 Modifications Introduced During Production

As with most of the German Panzers produced over an extended period during the War, during the production run of the Panzer IV/70 (A) a series of modifications were made to improve its performance or simplify production. The following is a list of significant modifications associated with the month in which they were introduced:

5.4.1 September 1944

Four steel-tyred, rubber saving roadwheels were mounted on the front two stations on each side due to the vehicle being front heavy.

Drahtgeflecht-Schürzen (wire mesh) replaced the solid soft steel aprons hanging on the hull sides on or about 18 September 1944.

Zimmerit, an anti-magnetic coating, was no longer applied to the surface of the armour due to the rumour that this coating caught fire as a result of hits from armour-piercing rounds.

5.4.2 December 1944

A new type of rear tow bracket was to be mounted at the centre of the lower hull rear to allow tow bars to be easily connected to a Bergepanzer.

The hull side armour plates were extended fore and aft and drilled to create towing brackets at each corner of the hull. The older style towing brackets attached with six bolts were frequently broken when a vehicle was stuck or when the slack was not taken out of the cables prior to attempting to tow broken down Panzer IV/70 (A).

The number of return rollers on each side was reduced from four to three to simplify production and save on bearings. Panzer IV/70 (A) Fgst.Nr.120539 completed in January 1945 still had four return rollers, but others off the assembly line in December 1944 had three return rollers.

From a list of weapons for each armoured vehicle dated 15 November 1944, the Panzer IV Lang (A) was to be outfitted with an M.G.42, M.P.40, and a Sturmgewehr 44. A Kugellager (ball mount) with a periscope and a 90° gebogene Lauf (bent barrel) was to be installed on the loader's hatch to be used for close defence. The hole in the loader's hatch lid for this mount was covered by a circular plate on Panzer IV/70 (A) Fgst.Nr.120539 completed in January 1945, but was present on many Panzer IV/70 (A) captured in 1945.

On 21 November 1944, a meeting about onboard weapons for all Panzer Fahrzeuge was held. Insights from the minutes show that new models of the Panther and Tiger, to appear in 1945, would be equipped with Sturmgewehr 44 and the Rundumfeuerlafette used on Sturmgeschütz and Jagdpanzer would be redesigned for the Sturmgewehr 44. But for the time being only Pz.IV Lang (A) is being equipped with the MP-Vorsatz Pz. (roof ball mount) as Erfahrungsbericht (experience reports) from the troops in the field are not yet available.

From a list of vision and sighting equipment dated 15 November 1944, one-third of the Panzer IV lang (A) were to be outfitted with a SF 14 Z (scissors periscope) and a Entfernungsmesser 0.9m (range finder). The three pegs for mounting this EM 0.9m are present on the superstructure roof of Panzer IV/70 (A) Fgst.Nr.120539 completed in January 1945.

Due to delays in production and the use of the most conveniently located components, last in, first out tendencies were created resulting in time lags of up to several months before an ordered modification was fully implemented on all production vehicles in a series. Thus a vehicle produced in January or February may not have had the modifications that should have taken place as early as December. During any stage in the War, they could not afford to delay production by back fitting old parts that were available nor could they afford to throw older parts away just because they did not meet the latest specifications.

5.5 Combat Service

With its increased firepower, the Panzer IV/70 (A) was originally intended to supplement units that had the Pz.Kpfw. IV. Thus, units would be created that had the more tactically flexible Pz.Kpfw.IV with the long range penetration power

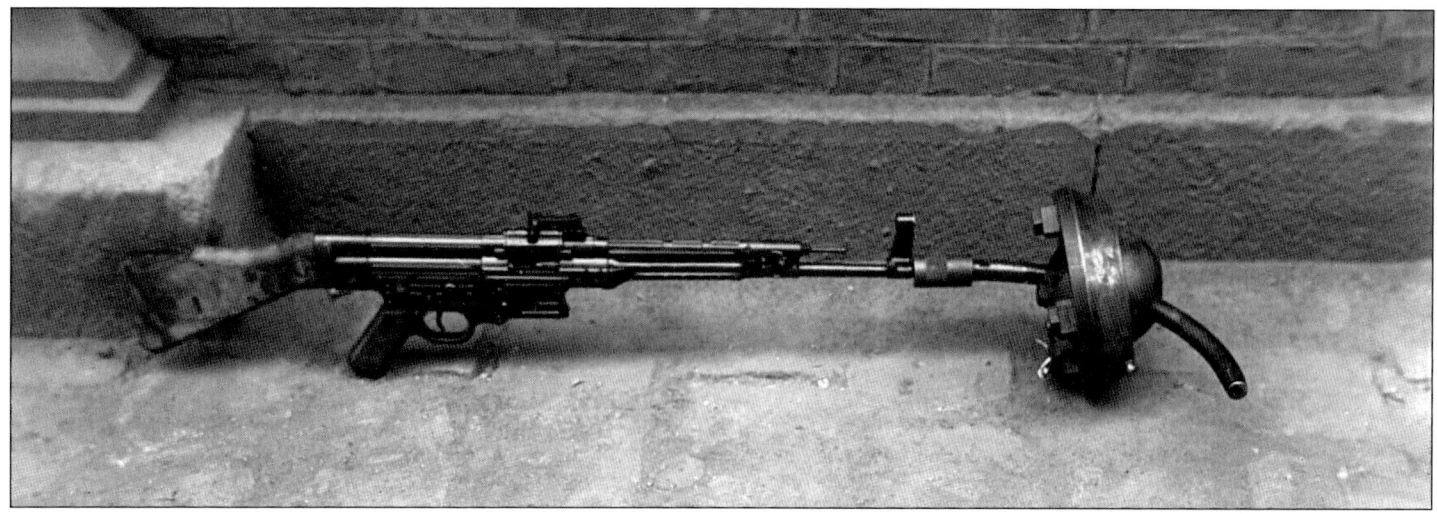

of the Panzer IV/70 (A). The plans for September 1944 were to send 68 Panzer IV/70 (A) as reinforcements to units that were already on the Eastern Front. In actual fact, only five were issued in September 1944. These went to the Panzer Kompanie of the Führer-Begleit-Brigade which also received 17 normal Pz.Kpfw.IV.

Due to production delays, only 17 were allocated to units on the Eastern Front in September, but were not actually shipped by rail until 15 October 1944. The remaining 51 of the original order for 68, were allocated to units on the Eastern Front in October with 17 being shipped by 18 October, 17 by 25 October and the last 17 on 30 October. These Panzer IV/70 (A) were eventually distributed to frontline units as follows:

No.	Regiment	Division
17	8./II./Pz.Rgt.6	3.Pz.Div.
4	9./III./Pz.Rgt.4	13.Pz.Div.
17	6./II./Pz.Rgt.39	17.Pz.Div.
13	9./III./Pz.Rgt.24	24.Pz.Div.
17	5./II./Pz.Rgt.9	25.Pz.Div.

In October 1944, the plans had called for 45 Panzer IV/70 (A) to be issued to the II.Abt./Pz.Rgt.Großdeutschland and 45 to the II.Abt./Pz.Rgt.2. No longer mixed with Pz.Kpfw. IV, the units were to consist of three companies with 14 Panzer IV/70 (A) with a further 3 Panzer IV/70 (A) as command vehicles for the Abteilung headquarters. Both of these units were to be utilised in the upcoming Ardennes offensive as independent army units attached to a division or corps other than their parent division.

Again due to production delays, the II.Abt./Pz.Rgt. Großdeutschland only received 38 Panzer IV/70 (A)

A Kugellager 'Vorsatz P' (ball mount) dismounted from a Panzer IV/70 (A) to allow the complete weapon to be demonstrated. The Sturmgewehr 44 (M.P.44) is attached to the 90° Krummerlauf (curved barrel) of the Vorsatz P. (ETO Tech. Intell. Rpt. 347, dated 18 July 1945). (NARA)

supplemented by 7 Pz.Kpfw.IV with all of them arriving at the unit by 4 December 1944 in time for the start of the Ardennes offensive.

The II.Abt./Pz.Rgt.2 was only issued 11 Panzer IV/70 (A) which had arrived at the unit on 16 November 1944. In November this unit also received 6 Pz.Kpfw.IV along with two companies with 22 Nashorn to fill out their complement. Shipped to Heeres Gruppe G on the Western Front, the II.Abt./Pz.Rgt.2 was available but did not take part in the Ardennes offensive.

In December a newly created unit, the Panzer Abteilung 208, was issued 31 normal Pz.Kpfw.IV and 14 Panzer IV/70 (A). Headquarters was outfitted with 3 Pz.Kpfw.IV, one company had all 14 Panzer IV/70 (A) and the other two companies each had 14 Pz.Kpfw.IV. In January 1945, Pz.Abt.208 was sent by rail to the Eastern Front and placed under the command of Heeres Gruppe Süd.

The only other issue of Panzer IV/70 (A) in December was 10 sent as replacements to the II.Abt./Pz.Rgt.25 of the 7.Panzer-Division in Heeres Gruppe Mitte on the Eastern Front.

Only two Panzer units received Panzer IV/70 (A) in January. The III.Abt./Pz.Rgt.24 of the 24.Panzer-Division under Heeres Gruppe Süd received 14 on 15 January 1945 and the I.Abt/Pz.Rgt.29 attached to Panzer Brigade 103 in Heeres Gruppe Mitte was issued 14 in January which arrived by rail on 2 February 1945. These were the last two Panzer units to be issued Panzer IV/70 (A).

From here on only Sturmgeschütz units were issued these Jagdpanzers. The decision was made to beef up the Sturmgeschütz units with a platoon of Panzer IV/70 (A)

Missing several pieces such as the external travel lock and armour cover for the machinegun port, the only surviving Panzer IV/70 (A), Fgst.Nr.120539, completed at Nibelungenwerk in January 1945, is on display at the Musée des Blindés in Saumur, France. (LA)

in order to provide long range anti-tank support. This was due to the increasing numbers of Shermans with 76mm and 17 pounder guns in the West and T34/85 and JS122 in the East. To train the crews, the Sturmgeschütz Schule at Burg received two Panzer IV/70 (A) in mid January. Issue to the units proceeded as follows:

No.	Unit	Arrived
4	Stu.G.Brig.244	26Jan45
3	Stu.G.Brig.341	26Jan45
3	Stu.G.Brig.394	26Jan45
3	Stu.G.Brig.902	26Jan45
3	Stu.G.Brig.280	13Feb45
3	Sturm Artl.Brig.905	13Feb45
3	Sturm Artl.Brig.911	13Feb45
3	Sturm Artl.Brig.667	22Mar45
3	Stu.G.Brig.243	4Feb45
3	Sturm Artl.Brig.236	8Feb45
3	Stu.G.Brig.301	8Feb45
31	Stu.G.Brig.G.D.	12Feb45
4	Stu.G.Brig.300	13Mar45
4	Stu.G.Brig.311	13Mar45
4	Stu.G.Brig.210	14Mar45
3	Stu.G.Brig.190	22Mar45
3	Stu.G.Brig.276	22Mar45
16	Sturm-Artl.Lehr Brig.111	After 15Mar45

Instead of only receiving a single platoon, the entire Stu.G.Brig.G.D. (Großdeutschland) was outfitted with three batteries of Panzer IV/70 (A) and Sturm Artillerie Lehr Brigade 111, the last unit to be given Panzer IV/70 (A), received a sufficient number to outfit four platoons.

The following after action account by a unit using the Panzer IV/70 (A) was written as a justification for the award of a Ritterkreuz to Leutnant Hartman of the 3.Batterie/Sturmgeschütz Brigade 311.

On 18 April 1945, once more a strong artillery barrage was started by the Russians. Leutnant Hartman drove forward to recon and determined that the Russian tanks had advanced over the railway embankment. He alerted his three Sturmgeschütz and drove back in his Fiat-Sportwagen. Two other Sturmgeschütz had been positioned during the night at the point of enemy penetration. Had they been knocked out by the enemy?

The three Sturmgeschütz (with Hartman in the Panzer IV/70 (A)) rolled through the railway underpass to the Obertor railway station and were already subjected to enemy artillery and mortar fire. Shortly thereafter, Hartman observed through his Schere (scissors periscope) several enemy JSU152. He opened fire and set the first on fire. Continually more of these steel colossi appeared. Hartman fired all of the Panzergranaten (armour-piercing rounds) at them and destroyed five of these giants. Both of his accompanying Sturmgeschütz took part in this attack. They likewise produced kills.

As Hartman drove back to reload ammunition, he discovered both of the Sturmgeschütz had been on night guard and was relieved that they were okay. In the interim, the two accompanying Sturmgeschütz had also destroyed five JSU152.

After loading ammunition, all of the Sturmgeschütz again advanced to attack. By itself, the Panzer IV/70 (A) of Hartman shot up 13 enemy tanks and JSU152. Altogether the enemy lost 25 armoured fighting vehicles and did not reach their objective which had been the Benderplatz on the other side of the Oder River in Breslau. If the enemy had obtained this objective, the entire island would have been lost and the German army would not have been able to hold the city of Breslau.

This account obviously relates a successful action and in no way reflects the average success of a Panzer IV/70 (A) against enemy armour. In most cases this late in the War the few operational Panzer IV/70 (A) in each unit would have been easily overwhelmed by the fully equipped opposing units on both the Eastern and Western Front. If they were not knocked out by frontal or flanking fire, they were forced to retire when the front lines were penetrated by enemy forces. The majority of them were lost due to mechanical breakdown or lack of fuel in positions where they could not be recovered due to rapidly advancing enemy forces.

Such was the fate of the only existing Panzer IV/70 (A) (Fahrgestell Nr.120539). It was knocked out by penetrations through the lower and upper superstructure on the right side by the machinegun port. These rounds were fired at close range from an American 75mm or 76mm gun by French Forces. This Panzer IV/70 (A) was captured in running condition by the 1st French Army and is at the Musée des Blindés at Saumur in France.

This Panzer IV/70 (A) was lost by the 6./Pz.Rgt.39, 17.Pz.Div. at Kobeřice, Czech Republic. It has three steel-tyred wheels and four return rollers. (VHA)

This completed Panzer IV/70 (A) at Nibelungenwerk has a raised base mount for the external travel lock and a full set of Drahtgeflechtschürzen (wire mesh Schürzen). It has been outfitted with a Kugellager (ball mount) and gebogene Lauf (curved barrel) for a M.P.44. (AMC)

Lacking components for their completion due to the bombing of Nibelungenwerk, these partially assembled Panzer IV/70 (A) were parked outside at the end of December 1944. Fahrgestell (chassis) with four steel-rimmed return rollers are mixed in with others that have three. The last two in the row already have the 7.5cm Pak 42 (L/70) and holders for Drahtgeflechtschürzen fitted. (WJS)

5.6 Panzer IV/70(A) Allocation

No.	Transported	Unit	Division	No.	Transported	Unit	Division
5	8Oct44	F.Begl.Brig.		3	18Jan45	StuG Brig.902	
7	15Oct44		1.Pz.Div.	14	24Jan45	I./Pz.Rgt.29	
10	15Oct44		13.Pz.Div.	3	28Jan45	StuG Brig.905	
17	18Oct44		3.Pz.Div.	3	28Jan45	StuG Brig.911	
17	25Oct44		17.Pz.Div.	3	28Jan45	StuG Brig.280	
17	30Oct44		25.Pz.Div.	3	3Feb45	StuG Brig.236	
11	9Nov44	II./Pz.Rgt.G.D.		3	3Feb45	StuG Brig.301	
11	13Nov44	II./Pz.Rgt.2		3	3Feb45	StuG Brig.667	
17	18Nov44	II./Pz.Rgt.G.D.		3	4Feb45	StuG Brig.243	
10	30Nov44	II./Pz.Rgt.G.D.		31	9Feb45	StuG Brig.G.D.	
14	19Dec44	Pz.Abt.208		4	7Mar45	StuG Brig.300	
10	2Jan45		7.Pz.Div.	4	7Mar45	StuG Brig.311	
14	13Jan45		24.Pz.Div.	4	7Mar45	StuG Brig.210	
3	18Jan45	StuG Brig.341		3	7Mar45	StuG Brig.190	
4	18Jan45	StuG Brig.244		3	7Mar45	StuG Brig.276	
3	18Jan45	StuG Brig.394		16	15Mar45	StuG Brig.111	

This Panzer IV/70 (A) Befehlswagen of 7./Pz.Rgt.2, Pz.Brigade.106 has an armour pot to protect the porcelain insulator below the base for a Sternantenne. The Befehlspanzer were outfitted with a Fu 8 long range radio set. The Vorsatz P Kugellager (ball mount) and gebogene Lauf (curved barrel) for a M.P.44 is mounted on the loader's hatch lid and there are pegs welded onto the superstructure roof for mounting an E.M. 0.9m rangefinder. (LA)

Starting in September 1944, Nibelungenwerk hung Drahtgeflechtschürzen on both sides of all Panzerkampfwagen IV and Panzer IV/70 (A). This Panzer IV/70 (A) was abandoned at Haus Neuglück, Bennerscheid near Bonn on 28 March 1945. The left side Drahtgeflechtschürzen was erroneously mounted backwards on the right side of this vehicle. (NARA)